Statistical handbook for non-statisticians

By the same author and available from McGraw-Hill Book Company (UK) Limited:

Elementary Analysis of Variance for the Behavioural Sciences

Statistical handbook
for non-statisticians

Ray Meddis

Bedford College,
London University

London · New York · St Louis · San Francisco · Düsseldorf · Johannesburg
Kuala Lumpur · Mexico · Montreal · New Delhi · Panama · Paris · São Paulo
Singapore · Sydney · Toronto

Published by

McGraw-Hill Book Company (UK) Limited
Maidenhead, Berkshire, England

Library of Congress Cataloging in Publication Data

Meddis, Ray.
    A statistical handbook for non-statisticians.

    Bibliography: p.158
    1. Statistics.   I.   Title.
HA29.M4388   519.5   74-14973
ISBN 0-07-084047-4
ISBN 0-07-084044-X (pbk.)

MADE AND PRINTED IN GREAT BRITAIN

Contents

Preface

There are two important stages in the learning of statistics. The first is the stage of acquisition which calls for lengthy and sympathetic explanations and good textbooks which move slowly and carefully across the new terrain. The second is the stage of application which calls for constant reminders of the thousands of minor details and manipulative procedures which are involved in even the simplest of statistical operations. At this stage, students need a reference book of formulae, procedures, tables and worked examples, a book which can be consulted quickly and easily. This handbook seeks to satisfy this need.

It is not a textbook but a *reference* book. It does not seek to teach but it does attempt to explain where the memory may have become confused. The book has been designed for easy reference and, as a result, should not be read from beginning to end. In it I have tried to collect together and make easily available many things which my own students have asked for. I have written it in the language which they appear to understand and which I hope you will understand.

The core of the book (chapters 4 to 8) is a summary of statistical tests. As far as possible they have been presented in a uniform pattern. Firstly, the procedure for carrying out the test is described. After this, the assumptions of the test are given along with an indication of what conclusions can be drawn from the result. This is followed by a worked example to answer queries about numerical manipulations involved. Where applicable the test is accompanied by tables of critical values for the statistics in question.

Wherever possible I have given the 'standard' procedure for each test. Most tests, however, allow for considerable variation in the procedure section. Wherever I was aware of a choice, I adopted what seemed to be the simplest procedure both to carry out and to understand. Sometimes I have taken the liberty of simplifying the standard procedure when I was aware of unnecessary computations. I hope that the simplifications I have made will prove a positive contribution.

The worked examples have been specifically designed to show up similarities between tests. Most two variable tests investigate the possibility that the two variables are related in some way. There are many different research designs, each requiring its own test, which can be used to ask the same question. By using the same basic example for different tests I hope to illustrate the underlying similarities as well as the important differences between tests. If this proves successful it should make you more aware of why you chose one test rather than another and why there need to be so very many tests to answer what is essentially the same question.

The tables in this book have all (except 7.1.2) been recalculated. This exercise was not, of course, necessary because existing tables are faulty in any way. It was necessitated by the desire to produce every table in the same format with critical values at the same four levels of statistical significance. As you will be aware uniformity in the style in which tables are presented is not a constraint which has concerned many authors of introductory texts. In my experience this produces considerable unnecessary confusion in the beginner and mature student alike. It is reassuring to note that my values agree closely with published critical values, whenever they are available at my chosen significance levels. Some small differences arise where I have chosen exact values which are slightly 'more significant' rather than a published value which is slightly 'less significant'. I have adopted throughout the book the principle of choosing exactly correct or slightly *conservative* values of the statistic in question for any given significance level.

It is appropriate at this point to defend briefly my choice of tests. Clearly, this book is not a compendium of all popular statistical tests. Rather it is a collection intended to cover all simple research *designs*. Wherever two or more tests apply equally well to the same design I have chosen to include only one. On one occasion where no appropriate test was immediately available (section 6.6) a stopgap has been proposed.

The choice of non-parametric tests was guided very much by Siegel's (1956) brilliant compilation. Tests not in Siegel's book were only included if Siegel did not supply an appropriate test for that design. This accounts for the presence of Page's L test (6.9) and a variant of the Kolmogorov–Smirnov procedures (4.5) as well as a variant of Kendall's correlation test (5.9). The choice of parametric tests was straightforward since they are all variants of analysis of variance or Pearson's product moment correlation coefficient.

Chapters 1, 2 and 3 are not directly concerned with statistical testing but supply a brief statement or *aide-memoire* in largely non-mathematical terms of many related topics. Chapter 3, for example, is devoted to the problems of choosing an appropriate test. If you have difficulty in deciding which test to use where, then you may well benefit from reading the whole chapter. Chapter 3 explains much of the vocabulary used in the book and will help you to 'tune in' to the method of presentation of the tests later in the book. Chapter 2 also describes symbols, formulae, conventions and standard procedures as well as the major frequency distributions (normal, Fisher's F, chi square, Student's t, and Poisson) which are constantly used for reference purposes.

Chapter 1 you will either need desperately or not at all. It gives help with arithmetic problems associated with handling the decimal point, logarithms, squares, square roots and simple probability calculations. I have attempted a unified approach to these topics using the 'characteristic–mantissa' approach. If you feel in need of assistance in this department, read the chapter carefully in its entirety and *practise*. Arithmetic is a skill not a fact!

References to original research papers are not given in this text. This is not intended as a discourtesy to the many statisticians who have contributed to the current state of the art. Rather it is an acknowledgement of the fact that the readers, for whom this book has been written, are unlikely ever to make use of this information. Indeed if they were to do so, it is quite possible that they would understand little of what they would find in these papers. Instead appendix B gives references to a small number of textbooks which give further information about the tests and themselves give the references to the original research. Wherever possible, I have tried to acknowledge our debt by retaining the test contributor's name (or names) in the title of the test.

Recalculation of the tables in this book required considerable use of computer facilities. Accordingly, I would like to acknowledge my gratitude to the staff of the computing unit at Bedford College. In particular, I must salute the patience and ingenuity of Richard Stephens who wrote many of the programs. His tact, cunning and apparently effortless style combined to produce results much more quickly than anyone could reasonably have expected. I must also thank my friend Ray Matthews for the careful and confident help in producing the Kolmogorov–Smirnov tables.

This book was made possible by my colleague John Valentine who has taught me statistics during coffee breaks for the last nine years. It has been a pleasure and a privilege to have received such expert individual tuition for so long a period of time. Careful analysis would show little in this book which did not derive directly or indirectly from his teaching. I only regret that my presentation is unlikely to do justice to his clarity of thinking. Hopefully, this book will reduce some of the needless frustration which is typically associated at beginners' level with statistical testing and leave the way free for more people to be infected by the enthusiasm for statistics which he has given me.

1. Numerical methods

1.1 Decimal point

The following operations require, for confident and successful execution, a high degree of skill in manipulating the decimal point : (a) multiplication, (b) division, (c) squares and square roots, (d) logarithms, (e) slide rule computations and (f) all probability calculations.

Mathematical scientists have developed a system of notation which renders most of these problems trivial. This system will be briefly outlined below and used in the exposition of various subjects for the rest of this section.

The basic feature of this system is that all numbers are broken down to the product two simpler numbers, e.g.,

$$364\,720 = 10^5 \times 3.6472$$

Since $10^5 = 10 \times 10 \times 10 \times 10 \times 10 = 100\,000$, it can be seen that the product of the two numbers on the right really does equal the number on the left.

(a) The second number of the pair always has a *single* digit to the left of the decimal point, e.g.,

$$3.6472$$
$$1.4$$
$$2.0$$

Such a number is sometimes called a *mantissa*.

(b) The first number of the pair is always the number 10 raised to a power, e.g.,

$$10^6, 10^3, 10^1, 10^0, 10^{-2}, 10^{-15}$$

The power that 10 is raised to (e.g., 6, 3, 1, 0, etc.) is sometimes called the *index* (sometimes the *characteristic*).

[It is more usual to write the mantissa first and the index second. The reverse order has been adopted here to maintain a notation consistent with logarithm notation which follows later in the section].

We find the index by noting how many places (to the left *or* right) we need to move the decimal point to convert the number to a mantissa, e.g.,

$$3647.2 = 10^3 \times 3.6472$$

We must move the decimal point three places *to the left* to produce a proper mantissa. Accordingly, our index is 3. When we move the decimal point along *to the right*, we make the index negative, e.g.,

$$0.000\,364\,72 = 10^{-4} \times 3.6472$$

Here, we had to move the decimal point four places to the right.

Note: When the number is already in mantissa form, the index is zero, e.g.,

$$4.32 = 10^0 \times 4.32$$

Of course, the value of 10^0 is 1 so that

$$10^0 \times 4.32 = 1 \times 4.32 = 4.32$$

The results of our calculations can be converted back to normal form simply by reversing the operation, e.g.,

$$10^{-6} \times 6.41 = 0.000\,0064\,1$$

$$10^{-6} \times 0.641 = 0.000\,000\,641$$

Here, we had to move the decimal point 6 places to the left, producing zero digits where necessary.

Negative signs do not influence the operation but are attached to the mantissa, e.g.,

$$-0.372 = 10^{-1} \times -3.72$$
$$-37.2 = 10^1 \times -3.72$$

The advantages of this notation are many. The most important advantage centres on the fact that the mantissa is always a value between 1 and 10. This makes the arithmetic much easier since we are familiar with such values. The index is used to keep a check on the decimal point throughout the calculations. It makes use of the important property:

$$10^x \times 10^y = 10^{(x+y)}$$

Thus, when multiplying two values of ten raised to a power, we need simply add the two powers together. Similarly, for division:

$$10^x / 10^y = 10^{(x-y)}$$

When dividing we merely subtract the second power.

Multiplication. When multiplying we deal with the mantissas and indices separately:

▶ Example. (a) 347×4763

$$347 \times 4763 = (10^2 \times 3.47) \times (10^3 \times 4.763)$$
$$= (10^2 \times 10^3) \times (3.47 \times 4.763)$$
$$= 10^5 \times 16.527\,61 = 1\,652\,761.0$$

(b) $0.347 \times 0.047\,63$

$$0.347 \times 0.047\,63 = (10^{-1} \times 3.47) \times (10^{-2} \times 4.763)$$
$$= (10^{-1} \times 10^{-2}) \times (3.47 \times 4.763)$$
$$= 10^{-3} \times 16.527\,61 = 0.016\,527\,1$$

Note. Care must be taken when adding negative indices, e.g.,

$$(-1) + (-2) = -3$$

Division. Again, the mantissas and indices are handled separately:

▶ Example. (a) $347/4763$

$$\frac{347}{4763} = \frac{10^2 \times 3.47}{10^3 \times 4.763} = \frac{10^2}{10^3} \times \frac{3.47}{4.763}$$
$$= 10^{2-3} \times 0.7285 = 10^{-1} \times 0.7285 = 0.072\,85$$

(b) $0.347/0.047\,63$

$$\frac{0.347}{0.047\,63} = \frac{10^{-1} \times 3.47}{10^{-2} \times 4.763} = \frac{10^{-1}}{10^{-2}} \times \frac{3.47}{4.763}$$
$$= 10^{(-1)-(-2)} \times 0.7285 = 10^1 \times 0.7285 = 7.285$$

Approximations. This notation system is peculiarly suitable to finding quick, approximate results. The procedure is to drop lesser significant figures *after* the conversion to mantissa index pairs.

▶ Example. (a)

$$0.347 \times 0.047\,63 = (10^{-1} \times 10^{-2}) \times (3.47 \times 4.763)$$

Now take approximate values for the mantissas which are easily multiplied.

$$= 10^{-3} \times (3.5 \times 5) \quad \text{approx.}$$
$$= 10^{-3} \times 17.5 \qquad \text{approx.}$$
$$= 0.0175 \qquad \text{approx.}$$

The exact answer (found above) is 0.0165 which is quite close to our approximation.

▶ Example. (b)

$$\frac{0.347}{0.04763} = \frac{10^{-1}}{10^{-2}} \times \frac{3.47}{4.763}$$

Now take approximate values for the mantissas

$$= 10^{(-1)-(-2)} \times \frac{3.5}{5} \quad \text{approx.}$$
$$= 10^{1} \times 0.7 \qquad \text{approx.}$$
$$= 7 \qquad \text{approx.}$$

The exact answer (found above) is 7.285.

Such approximate calculations require very little computational effort and illustrate the great facility with which the decimal point is kept in check. It should also be clear by now that slide rule computations which involve only figures between 1 and 10 are greatly assisted by this system of notation.

1.2 Logarithms

Table 1.2.1 gives logarithms for numbers between 1·00 and 9·99.

▶ Example.

$$\log(3) = 0.4771$$

$$\log(4.93) = 0.6928$$

These logarithms are called *common logarithms*.
Common logarithms have a base of 10 so that

$$3 = 10^{0.4771}$$

$$4.93 = 10^{0.6928}$$

Natural logarithms (with a base of 2·718 or 'e') are not given in this book but they may be calculated easily.

$$\log_e(x) = 2.303 \log_{10}(x)$$

▶ Example.

$$\log_e(3) = 2.303 \times 0.4771 = 1.099$$

Other logarithms can also be calculated using common logarithms

$$\log_q(x) = \frac{\log_{10}(x)}{\log_{10}(q)}$$

where q is the base of the logarithm in question.

Table 1.2.1   Common logarithms.

	0	0·01	0·02	0·03	0·04	0·05	0·06	0·07	0·08	0·09
1·0	0	0·0043	0·0086	0·0128	0·0170	0·0212	0·0253	0·0294	0·0334	0·0374
1·1	0·0414	0·0453	0·0492	0·0531	0·0569	0·0607	0·0645	0·0682	0·0719	0·0755
1·2	0·0792	0·0828	0·0864	0·0899	0·0934	0·0969	0·1004	0·1038	0·1072	0·1106
1·3	0·1139	0·1173	0·1206	0·1239	0·1271	0·1303	0·1335	0·1367	0·1399	0·1430
1·4	0·1461	0·1492	0·1523	0·1553	0·1584	0·1614	0·1644	0·1673	0·1703	0·1732
1·5	0·1761	0·1790	0·1818	0·1847	0·1875	0·1903	0·1931	0·1959	0·1987	0·2014
1·6	0·2041	0·2068	0·2095	0·2122	0·2148	0·2175	0·2201	0·2227	0·2253	0·2279
1·7	0·2304	0·2330	0·2355	0·2380	0·2405	0·2430	0·2455	0·2480	0·2504	0·2529
1·8	0·2553	0·2577	0·2601	0·2625	0·2648	0·2672	0·2695	0·2718	0·2742	0·2765
1·9	0·2788	0·2810	0·2833	0·2856	0·2878	0·2900	0·2923	0·2945	0·2967	0·2989
2·0	0·3010	0·3032	0·3054	0·3075	0·3096	0·3118	0·3139	0·3160	0·3181	0·3201
2·1	0·3222	0·3243	0·3263	0·3284	0·3304	0·3324	0·3345	0·3365	0·3385	0·3404
2·2	0·3424	0·3444	0·3464	0·3483	0·3502	0·3522	0·3541	0·3560	0·3579	0·3598
2·3	0·3617	0·3636	0·3655	0·3674	0·3692	0·3711	0·3729	0·3747	0·3766	0·3784
2·4	0·3802	0·3820	0·3838	0·3856	0·3874	0·3892	0·3909	0·3927	0·3945	0·3962
2·5	0·3979	0·3997	0·4014	0·4031	0·4048	0·4065	0·4082	0·4099	0·4116	0·4133
2·6	0·4150	0·4166	0·4183	0·4200	0·4216	0·4232	0·4249	0·4265	0·4281	0·4298
2·7	0·4314	0·4330	0·4346	0·4362	0·4378	0·4393	0·4409	0·4425	0·4440	0·4456
2·8	0·4472	0·4487	0·4502	0·4518	0·4533	0·4548	0·4564	0·4579	0·4594	0·4609
2·9	0·4624	0·4639	0·4654	0·4669	0·4683	0·4698	0·4713	0·4728	0·4742	0·4757
3·0	0·4771	0·4786	0·4800	0·4814	0·4829	0·4843	0·4857	0·4871	0·4886	0·4900
3·1	0·4914	0·4928	0·4942	0·4955	0·4969	0·4983	0·4997	0·5011	0·5024	0·5038
3·2	0·5051	0·5065	0·5079	0·5092	0·5105	0·5119	0·5132	0·5145	0·5159	0·5172
3·3	0·5185	0·5198	0·5211	0·5224	0·5237	0·5250	0·5263	0·5276	0·5289	0·5302
3·4	0·5315	0·5328	0·5340	0·5353	0·5366	0·5378	0·5391	0·5403	0·5416	0·5428
3·5	0·5441	0·5453	0·5465	0·5478	0·5490	0·5502	0·5514	0·5527	0·5539	0·5551
3·6	0·5563	0·5575	0·5587	0·5599	0·5611	0·5623	0·5635	0·5647	0·5658	0·5670
3·7	0·5682	0·5694	0·5705	0·5717	0·5729	0·5740	0·5752	0·5763	0·5775	0·5786
3·8	0·5798	0·5809	0·5821	0·5832	0·5843	0·5855	0·5866	0·5877	0·5888	0·5899
3·9	0·5911	0·5922	0·5933	0·5944	0·5955	0·5966	0·5977	0·5988	0·5999	0·6010
4·0	0·6021	0·6031	0·6042	0·6053	0·6064	0·6075	0·6085	0·6096	0·6197	0·6117
4·1	0·6128	0·6138	0·6149	0·6160	0·6170	0·6180	0·6191	0·6201	0·6212	0·6222
4·2	0·6232	0·6243	0·6253	0·6263	0·6274	0·6284	0·6294	0·6304	0·6314	0·6325
4·3	0·6335	0·6345	0·6355	0·6365	0·6375	0·6385	0·6395	0·6405	0·6415	0·6425
4·4	0·6435	0·6444	0·6454	0·6464	0·6474	0·6484	0·6493	0·6503	0·6513	0·6522
4·5	0·6523	0·6542	0·6551	0·6561	0·6571	0·6580	0·6590	0·6599	0·6609	0·6618
4·6	0·6628	0·6637	0·6646	0·6656	0·6665	0·6675	0·6684	0·6693	0·6702	0·6712
4·7	0·6721	0·6730	0·6739	0·6749	0·6758	0·6767	0·6776	0·6785	0·6794	0·6803
4·8	0·6812	0·6821	0·6830	0·6839	0·6848	0·6857	0·6866	0·6875	0·6884	0·6893
4·9	0·6902	0·6911	0·6920	0·6928	0·6937	0·6946	0·6955	0·6964	0·6972	0·6981
5·0	0·6990	0·6998	0·7007	0·7016	0·7024	0·7033	0·7042	0·7050	0·7059	0·7067
5·1	0·7076	0·7084	0·7093	0·7101	0·7110	0·7118	0·7126	0·7135	0·7143	0·7152
5·2	0·7160	0·7168	0·7177	0·7185	0·7193	0·7202	0·7210	0·7218	0·7226	0·7235
5·3	0·7243	0·7251	0·7259	0·7267	0·7275	0·7284	0·7292	0·7300	0·7308	0·7316
5·4	0·7324	0·7332	0·7340	0·7348	0·7356	0·7364	0·7372	0·7380	0·7388	0·7396
5·5	0·7404	0·7412	0·7419	0·7427	0·7435	0·7443	0·7451	0·7459	0·7466	0·7474

Example.

$$\log_2(4) = \frac{\log_{10}(4)}{\log_{10}(2)} = \frac{0 \cdot 6021}{0 \cdot 3010} = 2 \cdot 000$$

Unless otherwise specified, all logarithms in this book have a base of 10 so that

$$\log(x) \equiv \log_{10}(x)$$

Table 1.2.1 can be used to find the logarithm of *any* number once the number has been broken up into a mantissa and index pair (see section 1.1). The mantissa is converted to a logarithm mantissa using the tables while the index is simply added on to the decimal part of the logarithm.

▶ Example.   Find log(37·4)

$$
\begin{array}{ccccc}
 & & \text{Index} & & \text{Mantissa} \\
37 \cdot 4 & = & 10^1 & \times & 3 \cdot 74
\end{array}
$$

$$\log(37 \cdot 4) \;=\; 1 \;+\; 0 \cdot 5729 \;=\; 1 \cdot 5729$$

▶ Example.   Find log(0·0476)

$$0 \cdot 0476 = 10^{-2} \times 4 \cdot 76$$

$$\log(0 \cdot 0476) = -2 + 0 \cdot 6776 = -1 \cdot 3224$$

Table 1.2.1—continued.

	0	0·01	0·02	0·03	0·04	0·05	0·06	0·07	0·08	0·09
5·6	0·7482	0·7490	0·7497	0·7505	0·7513	0·7520	0·7528	0·7536	0·7543	0·7551
5·7	0·7559	0·7566	0·7574	0·7582	0·7589	0·7597	0·7604	0·7612	0·7619	0·7627
5·8	0·7634	0·7642	0·7649	0·7657	0·7664	0·7672	0·7679	0·7686	0·7694	0·7701
5·9	0·7709	0·7716	0·7723	0·7731	0·7738	0·7745	0·7752	0·7760	0·7767	0·7774
6·0	0·7782	0·7789	0·7796	0·7803	0·7810	0·7818	0·7825	0·7832	0·7839	0·7846
6·1	0·7853	0·7860	0·7868	0·7875	0·7882	0·7889	0·7896	0·7903	0·7910	0·7917
6·2	0·7924	0·7931	0·7938	0·7945	0·7952	0·7959	0·7966	0·7973	0·7980	0·7987
6·3	0·7993	0·8000	0·8007	0·8014	0·8021	0·8028	0·8035	0·8041	0·8048	0·8055
6·4	0·8062	0·8069	0·8075	0·8082	0·8089	0·8096	0·8102	0·8109	0·8116	0·8122
6·5	0·8129	0·8136	0·8142	0·8149	0·8156	0·8162	0·8169	0·8176	0·8182	0·8189
6·6	0·8195	0·8202	0·8209	0·8215	0·8222	0·8228	0·8235	0·8241	0·8248	0·8254
6·7	0·8261	0·8267	0·8274	0·8280	0·8287	0·8293	0·8299	0·8306	0·8312	0·8319
6·8	0·8325	0·8331	0·8338	0·8344	0·8351	0·8357	0·8363	0·8370	0·8376	0·8382
6·9	0·8388	0·8395	0·8401	0·8407	0·8414	0·8420	0·8426	0·8432	0·8439	0·8445
7·0	0·8451	0·8457	0·8463	0·8470	0·8476	0·8482	0·8488	0·8494	0·8500	0·8506
7·1	0·8513	0·8519	0·8525	0·8531	0·8537	0·8543	0·8549	0·8555	0·8561	0·8567
7·2	0·8573	0·8579	0·8585	0·8591	0·8597	0·8603	0·8609	0·8615	0·8621	0·8627
7·3	0·8633	0·8639	0·8645	0·8651	0·8657	0·8663	0·8669	0·8675	0·8681	0·8686
7·4	0·8692	0·8698	0·8704	0·8710	0·8716	0·8722	0·8727	0·8733	0·8739	0·8745
7·5	0·8751	0·8756	0·8762	0·8768	0·8774	0·8779	0·8785	0·8791	0·8796	0·8802
7·6	0·8808	0·8814	0·8820	0·8825	0·8831	0·8837	0·8842	0·8848	0·8854	0·8859
7·7	0·8865	0·8871	0·8876	0·8882	0·8887	0·8893	0·8899	0·8904	0·8910	0·8915
7·8	0·8921	0·8927	0·8932	0·8938	0·8943	0·8949	0·8954	0·8960	0·8955	0·8971
7·9	0·8976	0·8982	0·8978	0·8993	0·8998	0·9004	0·9009	0·9015	0·9020	0·9025
8·0	0·9031	0·9036	0·9042	0·9047	0·9053	0·9058	0·9063	0·9069	0·9074	0·9079
8·1	0·9085	0·9090	0·9096	0·9101	0·9106	0·9112	0·9117	0·9122	0·9128	0·9133
8·2	0·9138	0·9143	0·9149	0·9154	0·9159	0·9165	0·9170	0·9175	0·9180	0·9186
8·3	0·9191	0·9196	0·9201	0·9206	0·9212	0·9217	0·9222	0·9227	0·9232	0·9238
8·4	0·9243	0·9248	0·9253	0·9258	0·9263	0·9269	0·9274	0·9279	0·9284	0·9289
8·5	0·9294	0·9299	0·9304	0·9309	0·9315	0·9320	0·9325	0·9330	0·9335	0·9340
8·6	0·9345	0·9350	0·9355	0·9360	0·9365	0·9370	0·9375	0·9380	0·9385	0·9390
8·7	0·9395	0·9400	0·9405	0·9410	0·9415	0·9420	0·9425	0·9430	0·9435	0·9440
8·8	0·9445	0·9450	0·9455	0·9460	0·9465	0·9469	0·9474	0·9479	0·9484	0·9489
8·9	0·9494	0·9499	0·9504	0·9509	0·9513	0·9518	0·9523	0·9528	0·9533	0·9538
9·0	0·9542	0·9547	0·9552	0·9557	0·9562	0·9566	0·9571	0·9576	0·9581	0·9586
9·1	0·9590	0·9595	0·9600	0·9605	0·9609	0·9614	0·9619	0·9624	0·9628	0·9633
9·2	0·9638	0·9643	0·9647	0·9652	0·9657	0·9661	0·9666	0·9671	0·9675	0·9680
9·3	0·9685	0·9689	0·9694	0·9699	0·9703	0·9708	0·9713	0·9717	0·9722	0·9727
9·4	0·9731	0·9736	0·9741	0·9745	0·9750	0·9754	0·9759	0·9763	0·9768	0·9773
9·5	0·9777	0·9782	0·9786	0·9791	0·9795	0·9800	0·9805	0·9809	0·9814	0·9818
9·6	0·9823	0·9827	0·9832	0·9836	0·9841	0·9845	0·9850	0·9854	0·9859	0·9863
9·7	0·9868	0·9872	0·9877	0·9881	0·9886	0·9890	0·9894	0·9899	0·9903	0·9908
9·8	0·9912	0·9917	0·9921	0·9926	0·9930	0·9934	0·9939	0·9943	0·9948	0·9952
9·9	0·9956	0·9961	0·9965	0·9969	0·9974	0·9978	0·9983	0·9987	0·9991	0·9996

Antilogarithms. Table 1.2.2 gives the antilogarithms of numbers between 0·0 and 0·9, e.g.,

$$\text{antilog}(0{\cdot}334) = 2{\cdot}16$$

The index is retained as a power of 10, e.g., $\text{antilog}(3{\cdot}334) = 10^3 \times 2{\cdot}16 = 2160$

Negative logarithms must be converted into a negative index and positive mantissa before table 1.2.2 can be used, e.g.,

$$\text{antilog}(-1{\cdot}3224) = \text{antilog}((-2) + 0{\cdot}6776)$$
$$= \text{antilog}((-2) + 0{\cdot}678) = 10^{-2} \times 4{\cdot}76$$
$$= 0{\cdot}0476$$

or

$$\text{antilog}(-0{\cdot}3030) = \text{antilog}((-1) + 0{\cdot}6970) = 10^{-1} \times 4{\cdot}98$$
$$= 0{\cdot}498$$

Note. Only 3 significant figures are offered in table 1.2.2. You must also round your logarithm value to 3 significant figures before using the table.

Table 1.2.2   Antilogarithms.

	0	0·001	0·002	0·003	0·004	0·005	0·006	0·007	0·008	0·009
0	1·00	1·00	1·00	1·01	1·01	1·01	1·01	1·02	1·02	1·02
0·01	1·02	1·03	1·03	1·03	1·03	1·04	1·04	1·04	1·04	1·04
0·02	1·05	1·05	1·05	1·05	1·06	1·06	1·06	1·06	1·07	1·07
0·03	1·07	1·07	1·08	1·08	1·08	1·08	1·09	1·09	1·09	1·09
0·04	1·10	1·10	1·10	1·10	1·11	1·11	1·11	1·11	1·12	1·12
0·05	1·12	1·12	1·13	1·13	1·13	1·14	1·14	1·14	1·14	1·15
0·06	1·15	1·15	1·15	1·16	1·16	1·16	1·16	1·17	1·17	1·17
0·07	1·17	1·18	1·18	1·18	1·19	1·19	1·19	1·19	1·20	1·20
0·08	1·20	1·21	1·21	1·21	1·21	1·22	1·22	1·22	1·22	1·23
0·09	1·23	1·23	1·24	1·24	1·24	1·24	1·25	1·25	1·25	1·26
0·10	1·26	1·26	1·26	1·27	1·27	1·27	1·28	1·28	1·28	1·29
0·11	1·29	1·29	1·29	1·30	1·30	1·30	1·31	1·31	1·31	1·32
0·12	1·32	1·32	1·32	1·33	1·33	1·33	1·34	1·34	1·34	1·35
0·13	1·35	1·35	1·36	1·36	1·36	1·36	1·37	1·37	1·37	1·38
0·14	1·38	1·38	1·39	1·39	1·39	1·40	1·40	1·40	1·41	1·41
0·15	1·41	1·42	1·42	1·42	1·43	1·43	1·43	1·44	1·44	1·44
0·16	1·45	1·45	1·45	1·46	1·46	1·46	1·47	1·47	1·47	1·48
0·17	1·48	1·48	1·49	1·49	1·49	1·50	1·50	1·50	1·51	1·51
0·18	1·51	1·52	1·52	1·52	1·53	1·53	1·53	1·54	1·54	1·55
0·19	1·55	1·55	1·56	1·56	1·56	1·57	1·57	1·57	1·58	1·58
0·20	1·58	1·59	1·59	1·60	1·60	1·60	1·61	1·61	1·61	1·62
0·21	1·62	1·63	1·63	1·64	1·64	1·64	1·64	1·65	1·65	1·66
0·22	1·66	1·66	1·67	1·67	1·67	1·68	1·68	1·69	1·69	1·69
0·23	1·70	1·70	1·71	1·71	1·71	1·72	1·72	1·73	1·73	1·73
0·24	1·74	1·74	1·75	1·75	1·75	1·76	1·76	1·77	1·77	1·77
0·25	1·78	1·78	1·79	1·79	1·79	1·80	1·80	1·81	1·81	1·82
0·26	1·82	1·82	1·83	1·83	1·84	1·84	1·85	1·85	1·85	1·86
0·27	1·86	1·87	1·87	1·87	1·88	1·88	1·89	1·89	1·90	1·90
0·28	1·91	1·91	1·91	1·92	1·92	1·93	1·93	1·94	1·94	1·95
0·29	1·95	1·95	1·96	1·96	1·97	1·97	1·98	1·98	1·99	1·99
0·30	2·00	2·00	2·00	2·01	2·01	2·02	2·02	2·03	2·03	2·04
0·31	2·04	2·05	2·05	2·06	2·06	2·07	2·07	2·07	2·08	2·08
0·32	2·09	2·09	2·10	2·10	2·11	2·11	2·12	2·12	2·13	2·13
0·33	2·14	2·14	2·15	2·15	2·16	2·16	2·17	2·17	2·18	2·18
0·34	2·19	2·19	2·20	2·20	2·21	2·21	2·22	2·22	2·23	2·23
0·35	2·24	2·24	2·25	2·25	2·26	2·26	2·27	2·28	2·28	2·29
0·36	2·29	2·30	2·30	2·31	2·31	2·32	2·32	2·33	2·33	2·34
0·37	2·34	2·35	2·36	2·36	2·37	2·37	2·38	2·38	2·39	2·39
0·38	2·40	2·40	2·41	2·42	2·42	2·43	2·43	2·44	2·44	2·45
0·39	2·45	2·46	2·47	2·47	2·48	2·48	2·49	2·49	2·50	2·51
0·40	2·51	2·52	2·52	2·53	2·54	2·54	2·55	2·55	2·56	2·56
0·41	2·57	2·58	2·58	2·59	2·59	2·60	2·61	2·61	2·62	2·62
0·42	2·63	2·64	2·64	2·65	2·65	2·66	2·67	2·67	2·68	2·69
0·43	2·69	2·70	2·70	2·71	2·72	2·72	2·73	2·74	2·74	2·75
0·44	2·75	2·76	2·77	2·77	2·78	2·79	2·79	2·80	2·81	2·81
0·45	2·82	2·82	2·83	2·84	2·84	2·85	2·86	2·86	2·87	2·88
0·46	2·88	2·89	2·90	2·90	2·91	2·92	2·92	2·93	2·94	2·94
0·47	2·95	2·96	2·96	2·97	2·98	2·99	2·99	3·00	3·01	3·01
0·48	3·02	3·03	3·03	3·04	3·05	3·05	3·06	3·07	2·08	3·08
0·49	3·09	3·10	3·10	3·11	3·12	3·13	3·13	3·14	3·15	3·16
0·50	3·16	3·17	3·18	3·18	3·19	3·20	3·21	3·21	3·22	3·23

Multiplication. We can multiply two numbers by adding their logarithms and finding the antilogarithm of the result.

▶ Example.   Calculate 39.4×183

$$39.4 = 10^1 \times 3.94: \quad \log(39.4) = 1.5955$$
$$183 = 10^2 \times 1.83: \quad \log(183) = \underline{2.2625}$$
$$\text{by adding} \quad 3.8580$$
$$\text{antilog} = 10^3 \times 7.21$$
$$\text{answer:} \quad 39.4 \times 183 = 7210 \quad \text{approx.}$$

Division.   Division requires that we *subtract* the logarithms.

▶ Example.   Calculate $4.69/0.087$

$$4.69 = 10^0 \times 4.69: \quad \log(4.69) \qquad\qquad\qquad 0.6712$$
$$0.087 = 10^{-2} \times 8.7: \quad \log(0.087) = (-2) \times 0.9395 = \underline{-1.0605}$$
$$\text{by subtracting} \qquad\qquad 1.7317$$

$$\text{antilog} = 10^1 \times 5.40$$

$$\text{answer:} \quad 4.69/0.087 = 54 \quad \text{approx.}$$

Table 1.2.2—continued.

	0	0.001	0.002	0.003	0.004	0.005	0.006	0.007	0.008	0.009
0.51	3.24	3.24	3.25	3.26	3.27	3.27	3.28	3.29	3.30	3.30
0.52	3.31	3.32	3.33	3.33	3.34	3.35	3.36	3.37	3.37	3.38
0.53	3.39	3.40	3.40	3.41	3.42	3.43	3.44	3.44	3.45	3.46
0.54	3.47	3.48	3.48	3.49	3.50	3.51	3.52	3.52	3.53	3.54
0.55	3.55	3.56	3.56	3.56	3.57	3.58	3.59	3.60	3.61	3.62
0.56	3.63	3.64	3.65	3.66	3.66	3.67	3.68	3.69	3.70	3.71
0.57	3.72	3.72	3.73	3.74	3.75	3.76	3.77	3.78	3.78	3.79
0.58	3.80	3.81	3.82	3.83	3.84	3.85	3.85	3.86	3.87	3.88
0.59	3.89	3.90	3.91	3.92	3.93	3.94	3.94	3.95	3.96	3.97
0.60	3.98	3.99	4.00	4.01	4.02	4.03	4.04	4.05	4.06	4.06
0.61	4.07	4.08	4.09	4.10	4.11	4.12	4.13	4.14	4.15	4.16
0.62	4.17	4.18	4.19	4.20	4.21	4.22	4.23	4.24	4.25	4.26
0.63	4.27	4.28	4.29	4.30	4.31	4.32	4.33	4.34	4.35	4.36
0.64	4.37	4.38	4.39	4.40	4.41	4.42	4.43	4.44	4.45	4.46
0.65	4.47	4.48	4.49	4.50	4.51	4.52	4.53	4.54	4.55	4.56
0.66	4.57	4.58	4.59	4.60	4.61	4.62	4.63	4.65	4.66	4.67
0.67	4.68	4.69	4.70	4.71	4.72	4.73	4.74	4.75	4.76	4.78
0.68	4.79	4.80	4.81	4.82	4.83	4.84	4.85	4.86	4.88	4.89
0.69	4.90	4.91	4.92	4.93	4.94	4.95	4.97	4.98	4.99	5.00
0.70	5.01	5.02	5.04	5.05	5.06	5.07	5.08	5.09	5.11	5.12
0.71	5.13	5.14	5.15	5.16	5.18	5.19	5.20	5.21	5.22	5.24
0.72	5.25	5.26	5.27	5.28	5.30	5.31	5.32	5.33	5.35	5.36
0.73	5.37	5.38	5.40	5.41	5.42	5.43	5.45	5.46	5.47	5.48
0.74	5.50	5.51	5.52	5.53	5.55	5.56	5.57	5.58	5.60	5.61
0.75	5.62	5.64	5.65	5.66	5.68	5.69	5.70	5.71	5.73	5.74
0.76	5.75	5.77	5.78	5.79	5.81	5.82	5.83	5.85	5.86	5.87
0.77	5.89	5.90	5.92	5.93	5.94	5.96	5.97	5.98	6.00	6.01
0.78	6.03	6.04	6.05	6.07	6.08	6.10	6.11	6.12	6.14	6.15
0.79	6.17	6.18	6.19	6.21	6.22	6.24	6.25	6.27	6.28	6.30
0.80	6.31	6.32	6.34	6.35	6.37	6.38	6.40	6.41	6.43	6.44
0.81	6.46	6.47	6.49	6.50	6.52	6.53	6.55	6.56	6.58	6.59
0.82	6.61	6.62	6.64	6.65	6.67	6.68	6.70	6.71	6.73	6.75
0.83	6.76	6.78	6.79	6.81	6.82	6.84	6.85	6.87	6.89	6.90
0.84	6.92	6.93	6.95	6.97	6.98	7.00	7.01	7.03	7.05	7.06
0.85	7.08	7.10	7.11	7.13	7.14	7.16	7.18	7.19	7.21	7.23
0.86	7.24	7.26	7.28	7.29	7.31	7.33	7.35	7.36	7.38	7.40
0.87	7.41	7.43	7.45	7.46	7.48	7.50	7.52	7.53	7.55	7.57
0.88	7.59	7.60	7.62	7.64	7.66	7.67	7.69	7.71	7.73	7.74
0.89	7.76	7.78	7.80	7.82	7.83	7.85	7.87	7.89	7.91	7.93
0.90	7.94	7.96	7.98	8.00	8.02	8.04	8.05	8.07	8.09	8.11
0.91	8.13	8.15	8.17	8.18	8.20	8.22	8.24	8.26	8.28	8.30
0.92	8.32	8.34	8.36	8.38	8.39	8.41	8.43	8.45	8.47	8.49
0.93	8.51	8.53	8.55	8.57	8.59	8.61	8.63	8.65	8.67	8.69
0.94	8.71	8.73	8.75	8.77	8.79	8.81	8.83	8.85	8.87	8.89
0.95	8.91	8.93	8.95	8.97	8.99	9.02	9.04	9.06	9.08	9.10
0.96	9.12	9.14	9.16	9.18	9.20	9.23	9.25	9.27	9.29	9.31
0.97	9.33	9.35	9.38	9.40	9.42	9.44	9.46	9.48	9.51	9.53
0.98	9.55	9.57	9.59	9.62	9.64	9.66	9.68	9.71	9.73	9.75
0.99	9.77	9.79	9.82	9.84	9.86	9.89	9.91	9.93	9.95	9.98

Powers. Raising a number to any power requires that we multiply the logarithm of the number by the power.

▶ Example.   Calculate 1370^2

$$1370 = 10^3 \times 1\cdot370: \quad \log(1370) = 3\cdot1367$$

$$\log(137^2) = 2 \times \log(1370) = 2 \times 3\cdot1367 = 6\cdot2734$$

$$\text{antilog}(6\cdot2734) = 10^6 \times 1\cdot87$$

$$\text{answer:} \quad 1370^2 = 1\,870\,000 \quad \text{approx.}$$

Fractional powers can be handled in the same way.

Square roots.   These are a special case of fractional powers. To find the square root, divide the logarithm by 2.

▶ Example.   Find the square root of $0\cdot101$

$$0\cdot101 = 10^{-1} \times 1\cdot01: \quad \log(0\cdot101) = (-1) + 0\cdot0043 = -0\cdot9957$$

$$\log\sqrt{0\cdot101} = \tfrac{1}{2}\log 0\cdot101 = \tfrac{1}{2} \times -0\cdot9957 = -0\cdot497\,85$$

$$\text{antilog} -0\cdot497\,85 = \text{antilog}((-1) + 0\cdot502) = 10^{-1} \times 3\cdot18 = 0\cdot318$$

$$\text{answer:} \quad \sqrt{0\cdot101} = 0\cdot318$$

Table 1.3.1   Squares.

	0	0·01	0·02	0·03	0·04	0·05	0·06	0·07	0·08	0·09
1·0	1·00	1·02	1·04	1·06	1·08	1·10	1·12	1·14	1·17	1·19
1·1	1·21	1·23	1·25	1·28	1·30	1·32	1·35	1·37	1·39	1·42
1·2	1·44	1·46	1·49	1·51	1·54	1·56	1·59	1·61	1·64	1·66
1·3	1·69	1·72	1·74	1·77	1·80	1·82	1·85	1·88	1·90	1·93
1·4	1·96	1·99	2·02	2·04	2·07	2·10	2·13	2·16	2·19	2·22
1·5	2·25	2·28	2·31	2·34	2·37	2·40	2·43	2·46	2·50	2·53
1·6	2·56	2·59	2·62	2·66	2·69	2·72	2·76	2·79	2·82	2·86
1·7	2·89	2·92	2·96	2·99	3·03	3·06	3·10	3·13	3·17	3·20
1·8	3·24	3·28	3·31	3·35	3·39	3·42	3·46	3·50	3·53	3·57
1·9	3·61	3·65	3·69	3·72	3·76	3·80	3·84	3·88	3·92	3·96
2·0	4·00	4·04	4·08	4·12	4·16	4·20	4·24	4·28	4·33	4·37
2·1	4·41	4·45	4·49	4·54	4·58	4·62	4·67	4·71	4·75	4·80
2·2	4·84	4·88	4·93	4·97	5·02	5·06	5·11	5·15	5·20	5·24
2·3	5·29	5·34	5·38	5·43	5·48	5·52	5·57	5·62	5·66	5·71
2·4	5·76	5·81	5·86	5·90	5·95	6·00	6·05	6·10	6·15	6·20
2·5	6·25	6·30	6·35	6·40	6·45	6·50	6·55	6·60	6·66	6·71
2·6	6·76	6·81	6·86	6·92	6·97	7·02	7·08	7·13	7·18	7·24
2·7	7·29	7·34	7·40	7·45	7·51	7·56	7·62	7·67	7·73	7·78
2·8	7·84	7·90	7·95	8·01	8·07	8·12	8·18	8·24	8·29	8·35
2·9	8·41	8·47	8·53	8·58	8·64	8·70	8·76	8·82	8·88	8·94
3·0	9·00	9·06	9·12	9·18	9·24	9·30	9·36	9·42	9·49	9·55
3·1	9·61	9·67	9·73	9·80	9·86	9·92	9·99	10·05	10·11	10·18
3·2	10·24	10·30	10·37	10·43	10·50	10·56	10·63	10·69	10·76	10·82
3·3	10·89	10·96	11·02	11·09	11·16	11·22	11·29	11·36	11·42	11·49
3·4	11·56	11·63	11·70	11·76	11·83	11·90	11·97	12·04	12·11	12·18
3·5	12·25	12·32	12·39	12·46	12·53	12·60	12·67	12·74	12·82	12·89
3·6	12·96	13·03	13·10	13·18	13·25	13·32	13·40	13·47	13·54	13·62
3·7	13·69	13·76	13·84	13·91	13·99	14·06	14·14	14·21	14·29	14·36
3·8	14·44	14·52	14·59	14·67	14·75	14·82	14·90	14·98	15·05	15·13
3·9	15·21	15·29	15·37	15·44	15·52	15·60	15·68	15·76	15·84	15·92
4·0	16·00	16·08	16·16	16·24	16·32	16·40	16·48	16·56	16·65	16·73
4·1	16·81	16·89	16·97	17·06	17·14	17·22	17·31	17·39	17·47	17·56
4·2	17·64	17·72	17·81	17·89	17·98	18·06	18·15	18·23	18·32	18·40
4·3	18·49	18·58	18·66	18·75	18·84	18·92	19·01	19·10	19·18	19·27
4·4	19·36	19·45	19·54	19·62	19·71	19·80	19·89	19·98	20·07	20·16
4·5	20·25	20·34	20·43	20·52	20·61	20·70	20·79	20·88	20·98	21·07
4·6	21·16	21·25	21·34	21·44	21·53	21·62	21·72	21·81	21·90	22·00
4·7	22·09	22·18	22·28	22·37	22·47	22·56	22·66	22·75	22·85	22·94
4·8	23·04	23·14	23·23	23·33	23·43	23·52	23·62	23·72	23·81	23·91
4·9	24·01	24·11	24·21	24·30	24·40	24·50	24·60	24·70	24·80	24·90
5·0	25·00	25·10	25·20	25·30	25·40	25·50	25·60	25·70	25·81	25·91
5·1	26·01	26·11	26·21	26·32	26·42	26·52	26·63	26·73	26·83	26·94
5·2	27·04	27·14	27·25	27·35	27·46	27·56	27·67	27·77	27·88	27·98
5·3	28·09	28·20	28·30	28·41	28·52	28·62	28·73	28·84	28·94	29·05
5·4	29·16	29·27	29·38	29·48	29·59	29·70	29·81	29·92	30·03	30·14
5·5	30·25	30·36	30·47	30·58	30·69	30·80	30·91	31·02	31·14	31·25

1.3 Squares and square roots

1.3.1 Squares

Table 1.3.1 gives the values of squares for numbers between 1 and 10. This table can be used for any number if it is first converted to index-mantissa notation, e.g.,

$$7210 = 10^3 \times 7{\cdot}21$$

We may find the square of the number by finding the squares of the index and mantissa separately.

$$7210^2 = (10^3 \times 7{\cdot}21^2)^2 = (10^3)^2 \times 7{\cdot}21^2$$

The square of the index is found by doubling the power to which 10 is raised:

$$(10^3)^2 = 10^{(3 \times 2)} = 10^6$$

The square of the mentissa is found from table 1.3.1 $7{\cdot}21^2 = 51{\cdot}98$ approx.

The result is now $7210^2 = 10^6 \times 51{\cdot}98 = 51\,980\,000$ approx.

▶ Example.   Find $0{\cdot}392^2$ $0{\cdot}392 = 10^{-1} \times 3{\cdot}92$

$$0{\cdot}392^2 = (10^{-1} \times 3{\cdot}92)^2$$

$$= 10^{-2} \times 15{\cdot}37 = 0{\cdot}1537 \quad \text{approx.}$$

Table 1.3.1—continued.

	0	0·01	0·02	0·03	0·04	0·05	0·06	0·07	0·08	0·09
5·6	31·36	31·47	31·58	31·70	31·81	31·92	32·04	32·15	32·26	32·68
5·7	32·49	32·60	32·72	32·83	32·95	33·06	33·18	33·29	33·41	33·52
5·8	33·64	33·76	33·87	33·99	34·11	34·22	34·34	34·46	34·57	34·69
5·9	34·81	34·93	35·05	35·16	35·28	35·40	35·52	35·64	35·76	35·88
6·0	36·00	36·12	36·24	36·36	36·48	36·60	36·72	36·84	36·97	37·09
6·1	37·21	37·33	37·45	37·58	37·70	37·82	37·95	38·07	38·19	38·32
6·2	38·44	38·56	38·69	38·81	38·94	39·06	39·19	39·31	39·44	39·56
6·3	39·69	39·82	39·94	40·07	40·20	40·32	40·45	40·58	40·70	40·83
6·4	40·96	41·09	41·22	41·34	41·47	41·60	41·73	41·86	41·99	42·12
6·5	42·25	42·38	42·51	42·64	42·77	42·90	43·03	43·16	43·39	43·43
6·6	43·56	43·69	43·82	43·96	44·09	44·22	44·36	44·49	44·62	44·76
6·7	44·89	45·02	45·16	45·29	45·43	45·56	45·70	45·83	45·97	46·10
6·8	46·24	46·38	46·51	46·65	46·79	46·92	47·06	47·20	47·33	47·47
6·9	47·61	47·75	47·89	48·02	48·16	48·30	48·44	48·58	48·72	48·86
7·0	49·00	49·14	49·28	49·42	49·56	49·70	49·84	49·98	50·13	50·27
7·1	50·41	50·55	50·69	50·84	50·98	51·12	51·27	51·41	51·55	51·70
7·2	51·84	51·98	52·13	52·27	52·42	52·56	52·71	52·85	53·00	53·14
7·3	53·29	53·44	53·58	53·73	53·88	54·02	54·17	54·32	54·46	54·61
7·4	54·76	54·91	55·06	55·20	55·35	55·50	55·65	55·80	55·95	56·10
7·5	56·25	56·40	56·55	56·70	56·85	57·00	57·15	57·30	57·46	57·61
7·6	57·76	57·91	58·06	58·22	58·37	58·52	58·68	58·83	58·98	59·14
7·7	59·29	59·44	59·60	59·75	59·91	60·06	60·22	60·37	60·53	60·68
7·8	60·84	61·00	61·15	61·31	61·47	61·62	61·78	61·94	62·09	62·25
7·9	62·41	62·57	62·73	62·88	63·04	63·20	63·36	63·52	63·68	63·84
8·0	64·00	64·16	64·32	64·48	64·64	64·80	64·96	65·12	65·29	65·45
8·1	65·61	65·77	65·93	66·10	66·26	66·42	66·59	66·75	66·91	67·08
8·2	67·24	67·40	67·57	67·73	67·90	68·06	68·23	68·39	68·56	68·72
8·3	68·89	69·06	69·22	69·39	69·56	69·72	69·89	70·06	70·22	70·39
8·4	70·56	70·73	70·90	71·06	71·23	71·40	71·57	71·74	71·91	72·08
8·5	72·25	72·42	72·59	72·76	72·93	73·10	73·27	73·44	73·62	73·79
8·6	73·96	74·13	74·30	74·48	74·65	74·82	75·00	75·17	75·34	75·52
8·7	75·69	75·86	76·04	76·21	76·39	76·56	76·74	76·91	77·09	77·26
8·8	77·44	77·62	77·79	77·97	78·15	78·32	78·50	78·68	78·85	79·03
8·9	79·21	79·39	79·57	79·74	79·92	80·10	80·28	80·46	80·64	80·82
9·0	81·00	81·18	81·36	81·54	81·72	81·90	82·08	82·26	82·45	82·63
9·1	82·81	82·99	83·17	83·36	83·54	83·72	82·91	84·09	84·27	84·46
9·2	84·64	84·82	85·01	85·19	85·38	85·56	85·75	85·93	86·12	86·30
9·3	86·49	86·68	86·86	87·05	87·24	87·42	87·61	87·80	87·98	88·17
9·4	88·36	88·55	88·74	88·92	89·11	89·30	89·49	89·68	89·87	90·06
9·5	90·25	90·44	90·63	90·82	91·01	91·20	91·39	91·58	91·78	91·97
9·6	92·16	92·35	92·54	92·74	92·93	93·12	93·32	93·51	93·70	93·90
9·7	94·09	94·28	94·48	94·67	94·87	95·06	95·26	95·45	95·65	95·84
9·8	96·04	96·24	96·43	96·63	96·83	97·02	97·22	97·42	97·61	97·81
9·9	98·01	98·21	98·41	98·60	98·80	99·00	99·20	99·40	99·60	99·80

1.3.2 Square roots

Table 1.3.2 gives square roots for values between 1 and 10.

Table 1.3.3 gives square roots for values between 10 and 100. To find a square root, begin by converting your number to index-mantissa notation and then find the square root of both parts separately.

▶ Example.   Find $\sqrt{23\,500}$

$$23\,500 = 10^4 \times 2 \cdot 35$$

$$\sqrt{23\,500} = \sqrt{10^4} \times \sqrt{2 \cdot 35}$$

The square root of the index portion is found by dividing by 2 the power to which 10 is raised

$$\sqrt{10^4} = 10^{(4/2)} = 10^2$$

The square root of the mantissa is found in table

$$\sqrt{2 \cdot 35} = 1 \cdot 533$$

The result is now

$$\sqrt{23\,500} = 10^2 \times 1 \cdot 533 = 153 \cdot 3$$

Table 1.3.2   Square roots (1 to 10).

	0	0·01	0·02	0·03	0·04	0·05	0·06	0·07	0·08	0·09
1·0	1·000	1·005	1·010	1·015	1·020	1·025	1·030	1·034	1·039	1·044
1·1	1·049	1·054	1·058	1·063	1·068	1·072	1·077	1·082	1·086	1·091
1·2	1·095	1·100	1·105	1·109	1·114	1·118	1·122	1·127	1·131	1·136
1·3	1·140	1·145	1·149	1·153	1·158	1·162	1·166	1·170	1·175	1·179
1·4	1·183	1·187	1·192	1·196	1·200	1·204	1·208	1·212	1·217	1·221
1·5	1·225	1·229	1·233	1·237	1·241	1·245	1·249	1·253	1·257	1·261
1·6	1·265	1·269	1·273	1·277	1·281	1·285	1·288	1·292	1·296	1·300
1·7	1·304	1·308	1·311	1·315	1·319	1·323	1·327	1·330	1·334	1·338
1·8	1·342	1·345	1·349	1·353	1·356	1·360	1·364	1·367	1·371	1·375
1·9	1·378	1·382	1·386	1·389	1·393	1·396	1·400	1·404	1·407	1·411
2·0	1·414	1·418	1·421	1·425	1·428	1·432	1·435	1·439	1·442	1·446
2·1	1·449	1·453	1·456	1·459	1·463	1·466	1·470	1·473	1·476	1·480
2·2	1·483	1·487	1·490	1·493	1·497	1·500	1·503	1·507	1·510	1·513
2·3	1·517	1·520	1·523	1·526	1·530	1·533	1·536	1·539	1·543	1·546
2·4	1·549	1·552	1·556	1·559	1·562	1·565	1·568	1·572	1·572	1·578
2·5	1·581	1·584	1·587	1·591	1·594	1·597	1·600	1·603	1·606	1·609
2·6	1·612	1·616	1·619	1·622	1·625	1·628	1·631	1·634	1·637	1·640
2·7	1·643	1·646	1·649	1·652	1·655	1·658	1·661	1·664	1·667	1·670
2·8	1·673	1·676	1·679	1·682	1·685	1·688	1·691	1·694	1·697	1·700
2·9	1·703	1·706	1·709	1·712	1·715	1·718	1·720	1·723	1·726	1·729
3·0	1·732	1·735	1·738	1·741	1·744	1·746	1·749	1·752	1·755	1·758
3·1	1·761	1·764	1·766	1·769	1·772	1·775	1·778	1·780	1·783	1·786
3·2	1·789	1·792	1·794	1·797	1·800	1·803	1·806	1·808	1·811	1·814
3·3	1·817	1·819	1·822	1·825	1·828	1·830	1·833	1·836	1·838	1·841
3·4	1·844	1·847	1·849	1·852	1·855	1·857	1·860	1·863	1·865	1·868
3·5	1·871	1·873	1·876	1·879	1·881	1·884	1·887	1·889	1·892	1·895
3·6	1·897	1·900	1·903	1·905	1·908	1·910	1·913	1·916	1·918	1·921
3·7	1·924	1·926	1·929	1·931	1·934	1·936	1·939	1·942	1·944	1·947
3·8	1·949	1·952	1·954	1·957	1·960	1·962	1·956	1·967	1·970	1·972
3·9	1·975	1·977	1·980	1·982	1·985	1·987	1·990	1·992	1·995	1·997
4·0	2·000	2·002	2·005	2·007	2·010	2·012	2·015	2·017	2·020	2·022
4·1	2·025	2·027	2·030	2·032	2·035	2·037	2·040	2·042	2·045	2·047
4·2	2·049	2·052	2·054	2·057	2·059	2·062	2·064	2·066	2·069	2·071
4·3	2·074	2·076	2·078	2·081	2·083	2·086	2·088	2·090	2·093	2·095
4·4	2·098	2·100	2·102	2·105	2·107	2·110	2·112	2·114	2·117	2·119
4·5	2·121	2·124	2·126	2·128	2·131	2·133	2·135	2·138	2·140	2·142
4·6	2·145	2·147	2·149	2·152	2·154	2·156	2·159	2·161	2·163	2·166
4·7	2·168	2·170	2·173	2·175	2·177	2·179	2·182	2·184	2·186	2·189
4·8	2·191	2·193	2·195	2·198	2·200	2·202	2·205	2·207	2·209	2·211
4·9	2·214	2·216	2·218	2·220	2·223	2·225	2·227	2·229	2·232	2·234
5·0	2·236	2·238	2·241	2·243	2·245	2·247	2·249	2·252	2·254	2·256
5·1	2·258	2·261	2·263	2·265	2·267	2·268	2·272	2·274	2·276	2·278
5·2	2·280	2·283	2·285	2·287	2·289	2·291	2·293	2·296	2·298	2·300
5·3	2·302	2·304	2·307	2·309	2·311	2·313	2·315	2·317	2·319	2·322
5·4	2·324	2·326	2·328	2·330	2·332	2·335	2·337	2·339	2·341	2·343
5·5	2·345	2·347	2·349	2·352	2·354	2·356	2·358	2·360	2·362	2·364
5·6	2·366	2·369	2·371	2·373	2·375	2·377	2·379	2·381	2·383	2·385

Because the index must remain a whole number, division by 2 may present a problem. When the index has an *odd* value, we *reduce* it by one (i.e., subtract 1) and move the decimal place of the mantissa one place to the *right* to compensate, e.g.,

$$3410 = 10^3 \times 3\cdot41$$
$$= 10^2 \times 34\cdot1$$

The square root of the index can now be found in the usual way. The square root of the mantissa can now be found in table 1.3.3 (values 10–100)

▶ Example.

$$\sqrt{3410} = \sqrt{10^2} \times \sqrt{34\cdot1}$$
$$= 10^{2/2} \times \sqrt{34\cdot1} = 10^1 \times 5\cdot84 = 58\cdot4$$

Table 1.3.2—continued.

	0	0·01	0·02	0·03	0·04	0·05	0·06	0·07	0·08	0·09
5·7	2·387	2·390	2·392	2·394	2·396	2·398	2·400	2·402	2·404	2·406
5·8	2·408	2·410	2·412	2·415	2·417	2·419	2·421	2·423	2·425	2·427
5·9	2·429	2·431	2·433	2·435	2·437	2·439	2·441	2·443	2·445	2·447
6·0	2·449	2·452	2·454	2·456	2·458	2·460	2·462	2·464	2·466	2·468
6·1	2·470	2·472	2·474	2·476	2·478	2·470	2·482	2·484	2·486	2·488
6·2	2·490	2·492	2·494	2·496	2·498	2·500	2·502	2·504	2·506	2·508
6·3	2·510	2·512	2·514	2·516	2·518	2·520	2·522	2·524	2·526	2·528
6·4	2·530	2·532	2·534	2·536	2·538	2·540	2·542	2·544	2·546	2·548
6·5	2·550	2·551	2·553	2·555	2·557	2·559	2·561	2·563	2·565	2·567
6·6	2·569	2·571	2·573	2·575	2·577	2·579	2·581	2·583	2·585	2·587
6·7	2·588	2·590	2·592	2·594	2·596	2·598	2·600	2·602	2·604	2·606
6·8	2·608	2·610	2·612	2·613	2·615	2·617	2·619	2·621	2·623	2·625
6·9	2·627	2·629	2·631	2·632	2·634	2·636	2·638	2·640	2·642	2·644
7·0	2·646	2·648	2·650	2·651	2·653	2·655	2·657	2·659	2·661	2·663
7·1	2·655	2·666	2·668	2·670	2·672	2·674	2·676	2·678	2·680	2·681
7·2	2·683	2·685	2·687	2·689	2·691	2·693	2·694	2·696	2·698	2·700
7·3	2·702	2·704	2·706	2·707	2·709	2·711	2·713	2·715	2·717	2·718
7·4	2·720	2·722	2·724	2·726	2·728	2·729	2·731	2·733	2·735	2·737
7·5	2·739	2·740	2·742	2·744	2·746	2·748	2·750	2·751	2·753	2·755
7·6	2·757	2·759	2·760	2·762	2·764	2·766	2·768	2·769	2·771	2·773
7·7	2·775	2·777	2·778	2·780	2·782	2·784	2·786	2·787	2·789	2·791
7·8	2·793	2·795	2·796	2·798	2·800	2·802	2·804	2·805	2·807	2·809
7·9	2·811	2·812	2·814	2·816	2·818	2·820	2·821	2·823	2·825	2·827
8·0	2·828	2·830	2·832	2·834	2·835	2·837	2·839	2·841	2·843	2·844
8·1	2·846	2·848	2·850	2·851	2·853	2·855	2·857	2·858	2·860	2·862
8·2	2·864	2·865	2·867	2·869	2·871	2·874	2·874	2·876	2·877	2·879
8·3	2·881	2·883	2·884	2·886	2·888	2·890	2·891	2·893	2·895	2·897
8·4	2·898	2·900	2·902	2·903	2·905	2·907	2·909	2·910	2·912	2·914
8·5	2·915	2·917	2·919	2·921	2·922	2·924	2·926	2·927	2·929	2·931
8·6	2·933	2·934	2·936	2·938	2·939	2·941	2·943	2·944	2·946	2·948
8·7	2·950	2·951	2·953	2·955	2·956	2·958	2·960	2·961	2·963	2·965
8·8	2·966	2·968	2·970	2·972	2·973	2·975	2·977	2·978	2·980	2·982
8·9	2·983	2·985	2·987	2·988	2·990	2·992	2·993	2·995	2·997	2·998
9·0	3·000	3·002	3·003	3·005	3·007	3·008	3·010	3·012	3·013	3·015
9·1	3·017	3·018	3·020	3·022	3·023	3·025	3·027	3·028	3·030	3·032
9·2	3·033	3·035	3·036	3·038	3·040	3·041	3·043	3·045	3·046	3·048
9·3	3·050	3·051	3·053	3·055	3·056	3·058	3·059	3·061	3·063	3·064
9·4	3·066	3·068	3·069	3·071	3·072	3·074	3·076	3·077	3·079	3·081
9·5	3·082	3·084	3·085	3·087	3·089	3·090	3·092	3·094	3·095	3·097
9·6	3·098	3·100	3·102	3·103	3·105	3·106	3·108	3·110	3·111	3·113
9·7	3·114	3·116	3·118	3·119	3·121	3·122	3·124	3·126	3·127	3·129
9·8	3·130	3·132	3·134	3·135	3·137	3·138	3·140	3·142	3·143	3·145
9·9	3·146	3·148	3·159	3·151	3·153	3·154	3·156	3·158	3·159	3·161
10·0	3·162	3·164	3·165	3·167	3·169	3·170	3·172	3·173	3·175	3·176

▶ Example. Calculate $\sqrt{0.0094}$

$$0.0094 = 10^{-3} \times 9.4$$

but -3 is odd and must be adjusted by subtracting 1

$$= 10^{-4} \times 94.0$$

$$\sqrt{0.0094} = \sqrt{10^{-4}} \times \sqrt{94} = 10^{-2} \times 9.695$$

answer: $\sqrt{0.0094} = 0.096\,95$

Table 1.3.3   Square roots (10 to 100).

	0	0.1	0.2	0.3	0.4	0.5	0.6	0.7	0.8	0.9
10	3·162	3·178	3·194	3·209	3·225	3·240	3·256	3·271	3·286	3·302
11	3·317	3·332	3·347	3·362	3·376	3·391	3·406	3·421	3·435	3·450
12	3·464	3·479	3·493	3·507	3·521	3·536	3·550	3·564	3·578	3·592
13	3·606	3·619	3·633	3·647	3·661	3·674	3·688	3·701	3·715	3·728
14	3·742	3·755	3·768	3·782	3·795	3·808	3·821	3·834	3·847	3·860
15	3·873	3·886	3·899	3·912	3·924	3·937	3·950	3·962	3·975	3·987
16	4·000	4·012	4·025	4·037	4·050	4·062	4·074	4·087	4·099	4·111
17	4·123	4·135	4·147	4·159	4·171	4·183	4·195	4·207	4·219	4·231
18	4·243	4·254	4·266	4·278	4·290	4·301	4·313	4·324	4·336	4·347
19	4·359	4·370	4·382	4·393	4·405	4·416	4·427	4·438	4·450	4·461
20	4·472	4·483	4·494	4·506	4·517	4·528	4·539	4·550	4·561	4·572
21	4·583	4·593	4·604	4·615	4·626	4·637	4·648	4·658	4·669	4·680
22	4·690	4·701	4·712	4·722	4·733	4·743	4·754	4·764	4·775	4·785
23	4·796	4·806	4·817	4·827	4·837	4·848	4·858	4·868	4·879	4·889
24	4·899	4·909	4·919	4·930	4·940	4·950	4·960	4·970	4·980	4·990
25	5·000	5·010	5·020	5·030	5·040	5·050	5·060	5·070	5·079	5·089
26	5·099	5·109	5·119	5·128	5·138	5·148	5·158	5·167	5·177	5·187
27	5·196	5·206	5·215	5·225	5·235	5·244	5·254	5·263	5·273	5·282
28	5·292	5·301	5·310	5·320	5·329	5·339	5·348	5·357	5·367	5·376
29	5·385	5·394	5·404	5·413	5·422	5·431	5·441	5·450	5·459	5·468
30	5·477	5·486	5·495	5·505	5·514	5·523	5·532	5·541	5·550	5·559
31	5·568	5·577	5·586	5·595	5·604	5·612	5·621	5·630	5·639	5·648
32	5·657	5·666	5·675	5·683	5·692	5·701	5·710	5·718	5·727	5·736
33	5·745	5·753	5·762	5·771	5·779	5·788	5·797	5·805	5·814	5·822
34	5·831	5·840	5·848	5·857	5·865	5·874	5·882	5·891	5·899	5·908
35	5·916	5·925	5·933	5·941	5·950	5·958	5·967	5·975	5·983	5·992
36	6·000	6·008	6·017	6·025	6·033	6·042	6·050	6·058	6·066	6·075
37	6·083	6·091	6·099	6·107	6·116	6·124	6·132	6·140	6·148	6·156
38	6·164	6·173	6·181	6·189	6·197	6·205	6·213	6·221	6·229	6·237
39	6·245	6·253	6·261	6·269	6·277	6·285	6·293	6·301	6·309	6·317
40	6·325	6·332	6·340	6·348	6·356	6·364	6·372	6·380	6·387	6·395
41	6·403	6·411	6·419	6·427	6·434	6·442	6·450	6·458	6·465	6·473
42	6·481	6·488	6·496	6·504	6·512	6·519	6·527	6·535	6·542	6·550
43	6·557	6·565	6·573	6·580	6·588	6·595	6·603	6·611	6·618	6·626
44	6·633	6·641	6·648	6·656	6·663	6·671	6·678	6·686	6·693	6·701
45	6·708	6·716	6·723	6·731	6·738	6·745	6·753	6·760	6·768	6·775
46	6·782	6·790	6·797	6·804	6·812	6·819	6·826	6·834	6·841	6·848
47	6·856	6·863	6·870	6·877	6·885	6·892	6·899	6·907	6·914	6·921
48	6·928	6·935	6·943	6·950	6·957	6·964	6·971	6·979	6·986	6·993
49	7·000	7·007	7·014	7·021	7·029	7·036	7·043	7·050	7·057	7·064
50	7·071	7·078	7·085	7·092	7·099	7·106	7·113	7·120	7·127	7·134
51	7·141	7·148	7·155	7·162	7·169	7·176	7·183	7·190	7·197	7·204
52	7·211	7·218	7·225	7·232	7·239	7·246	7·253	7·259	7·266	7·273
53	7·280	7·287	7·294	7·301	7·308	7·314	7·321	7·328	7·335	7·342
54	7·348	7·355	7·362	7·369	7·376	7·382	7·389	7·396	7·403	7·409
55	7·416	7·423	7·430	7·436	7·443	7·450	7·457	7·463	7·470	7·477

Table 1.3.3—continued.

	0	0·1	0·2	0·3	0·4	0·5	0·6	0·7	0·8	0·9
56	7·483	7·490	7·497	7·503	7·510	7·517	7·523	7·530	7·537	7·543
57	7·550	7·556	7·563	7·570	7·576	7·583	7·589	7·596	7·603	7·609
58	7·616	7·622	7·629	7·635	7·642	7·649	7·655	7·662	7·668	7·675
59	7·681	7·688	7·694	7·701	7·707	7·714	7·720	7·727	7·733	7·740
60	7·746	7·752	7·759	7·765	7·772	7·778	7·785	7·791	7·797	7·804
61	7·810	7·817	7·823	7·829	7·836	7·842	7·849	7·855	7·861	7·868
62	7·874	7·880	7·887	7·893	7·899	7·906	7·912	7·918	7·925	7·931
63	7·937	7·944	7·950	7·956	7·962	7·969	7·975	7·981	7·987	7·994
64	8·000	8·006	8·012	8·019	8·025	8·031	8·037	8·044	8·050	8·056
65	8·062	8·068	8·075	8·081	8·087	8·093	8·099	8·106	8·112	8·118
66	8·124	8·130	8·136	8·142	8·149	8·155	8·161	8·167	8·173	8·179
67	8·185	8·191	8·198	8·204	8·210	8·216	8·222	8·228	8·234	8·240
68	8·246	8·252	8·258	8·264	8·270	7·276	8·283	8·289	8·295	8·301
69	8·307	8·313	8·319	8·325	8·331	8·337	8·343	8·349	8·355	8·361
70	8·367	8·373	8·379	8·385	8·390	8·396	8·402	8·408	8·414	8·420
71	8·426	8·432	8·438	8·444	8·450	8·456	8·462	8·468	8·473	8·479
72	8·485	8·491	8·497	8·503	8·509	8·515	8·521	8·526	8·532	8·538
73	8·544	8·550	8·556	8·562	8·567	8·573	8·579	8·585	8·591	8·597
74	8·602	8·608	8·614	8·620	8·626	8·631	8·627	8·643	8·649	8·654
75	8·660	8·666	8·672	8·678	8·683	8·689	8·695	8·701	8·706	8·712
76	8·718	8·724	8·729	8·735	8·741	8·746	8·752	8·758	8·764	8·769
77	8·775	8·781	8·786	8·792	8·798	8·803	8·809	8·815	8·820	8·826
78	8·832	8·837	8·843	8·849	8·854	8·860	8·866	8·871	8·877	8·883
79	8·888	8·894	8·899	8·905	8·911	8·916	8·922	8·927	8·933	8·939
80	8·944	8·950	8·955	8·961	8·967	8·972	8·978	8·983	8·989	8·994
81	9·000	9·006	9·011	9·017	9·022	9·028	9·033	9·039	9·044	9·050
82	9·055	9·061	9·066	9·072	9·077	9·083	9·088	9·094	9·099	9·105
83	9·110	9·116	9·121	9·127	9·132	9·138	9·143	9·149	9·154	9·160
84	9·165	9·171	9·176	9·182	9·187	9·192	9·198	9·203	9·209	9·214
85	9·220	9·225	9·230	9·236	9·241	9·247	9·252	9·257	9·263	9·268
86	9·274	9·279	9·284	9·290	9·295	9·301	9·306	9·311	9·317	9·322
87	9·327	9·333	9·337	9·343	9·349	9·354	9·359	9·365	9·370	9·375
88	9·381	9·386	9·391	9·397	9·402	9·407	9·413	9·418	9·423	9·429
89	9·434	9·439	9·445	9·450	9·455	9·460	9·466	9·471	9·476	9·482
90	9·487	9·492	9·497	9·503	9·508	9·513	9·518	9·524	9·529	9·534
91	9·539	9·545	9·550	9·555	9·560	9·566	9·571	9·576	9·581	9·586
92	9·592	9·597	9·602	9·607	9·612	9·618	9·623	9·628	9·633	9·638
93	9·644	9·649	9·654	9·659	9·664	9·670	9·675	9·680	9·685	9·690
94	9·695	9·701	9·706	9·711	9·716	9·721	9·726	9·731	9·737	9·742
95	9·747	9·752	9·757	9·762	9·767	9·772	9·778	9·783	9·788	9·793
96	9·798	9·803	9·808	9·813	9·818	9·823	9·829	9·834	9·839	9·844
97	9·849	9·854	9·859	9·864	9·869	9·874	9·879	9·884	9·889	9·894
98	9·899	9·905	9·915	9·915	9·920	9·925	9·930	9·935	9·940	9·945
99	9·950	9·955	9·960	9·965	9·970	9·975	9·980	9·985	9·990	9·995
100	10·000									

1.4 Probability

1.4.1 *A priori* probability

To evaluate in advance the probability that a given type of event will occur, we need to be able to analyse the possible outcomes into a set of *equally likely* events.

If this is possible we note (a) the total number of possible outcomes—n, and (b) the total number of possible outcomes which satisfy the description of the type of event we are interested in—m. The probability that one of the m possibilities will occur is p where:

$$p = \frac{m}{n} \qquad (1.4.1)$$

▶ Example.   When throwing a die what is the probability of having an odd number face up?

$$n = \text{total number of possible outcomes} \quad 6$$

$$m = \text{total number of odd faces} \qquad\qquad 3$$

$$p = \frac{m}{n} = \frac{3}{6} = 0.5$$

The probability of throwing an odd is 0.5.

1.4.2 Permutations and factorials

If there are n objects, there are $n!$ ways of ordering them.

$$n! \text{ is read '}n \text{ factorial'}$$

$$n! = n \times (n - 1) \times (n - 2) \times \ldots \times 3 \times 2 \times 1 \qquad (1.4.2)$$

Table 1.4 gives values of $n!$ for $n = 1$ to 18 and of $\log(n!)$ for $n = 1$ to 100.

▶ Example.   How many ways can we order the letters A B C.
There are 3 objects (letters); $n = 3$.
There are 3! ways of ordering them

$$3! = 3 \times 2 \times 1 = 6$$

They are in fact

$$\text{A B C} \qquad \text{B A C} \qquad \text{C A B}$$

$$\text{A C B} \qquad \text{B C A} \qquad \text{C B A}$$

▶ Example.   In a competition, the entrant has to specify exactly a judge's order of preference for 6 moral virtues. If the entrant chooses his suggested order at random, how likely is it that he will guess correctly?
   There are 6! ways of ordering the moral virtues. From table 1.4: 6! = 720.
   Only one ordering is the winning permutation thus

$$P = \frac{1}{720} = 0.0014$$

The probability of guessing the correct reply is 0.0014.

Combinations.   There are $_nC_r$ ways of choosing a set of r objects from a total collection of n objects where

$$_nC_r = \frac{n!}{r!(n - r)!} \qquad (1.4.3)$$

$_nC_r$ is sometimes written $\binom{n}{r}$

▶ Example.   How many different triplets of names can be drawn from this collection?

$$\text{Rabbit} \qquad \text{Hare} \qquad \text{Squirrel} \qquad \text{Hedgehog} \qquad \text{Badger}$$

There are 5 names: $n = 5$.

Table 1.4 Factorials and log factorials.

N	$\log_{10}(N!)$	$N!$
1	0	1·0
2	0·301 03	2·0
3	0·778 15	6·0
4	1·380 21	24·0
5	2·079 18	120·0
6	2·857 33	720·0
7	3·702 43	4 040·0
8	4·605 52	40 320·0
9	5·559 76	362 880·0
10	6·559 76	3 628 800·0
11	7·601 16	39 916 800·0
12	8·680 34	479 001·600·0
13	9·794 28	6 227 020·800·0
14	10·940 41	87 178 291 200·0
15	12·116 50	1 307 674 368 000·0
16	13·320 62	20 922 789 888 000·0
17	14·551 07	355 687 428 096 000·0
18	15·806 34	6 402 373 705 728 000·0
19	17·085 09	
20	18·386 12	
21	19·708 34	
22	21·050 77	
23	22·412 49	
24	23·792 71	
25	25·190 65	
26	26·605 62	
27	28·036 98	
28	29·484 14	
29	30·946 54	
30	32·423 66	
31	33·915 02	
32	35·420 17	
33	36·938 69	
34	38·470 16	
35	40·014 23	
36	41·570 54	
37	43·138 74	
38	44·718 52	
39	46·309 59	
40	47·911 65	
41	49·524 43	
42	51·147 68	
43	52·781 15	
44	54·424 60	
45	56·077 81	
46	57·740 57	
47	59·412 67	
48	61·093 91	
49	62·784 10	
50	64·483 07	
51	66·190 65	
52	67·906 65	
53	69·630 92	

N	$\log_{10}(N!)$	N	$\log_{10}(N!)$
54	71·363 32	81	120·763 21
55	73·103 68	82	122·677 03
56	74·851 87	83	124·596 10
57	76·607 74	84	126·520 38
58	78·371 17	85	128·449 80
59	80·142 02	86	130·384 30
60	81·920 17	87	132·323 82
61	83·705 50	88	134·268 30
62	85·497 90	89	136·217 69
63	87·297 24	90	138·171 94
64	89·103 42	91	140·130 98
65	90·916 33	92	142·094 77
66	92·735 87	93	144·063 25
67	94·561 95	94	146·036 38
68	96·394 46	95	148·014 10
69	98·233 31	96	149·996 37
70	100·078 41	97	151·983 14
71	101·929 66	98	153·974 37
72	103·787 00	99	155·970 00
73	105·650 32	100	157·970 00
74	107·519 55		
75	109·394 61		
76	111·275 43		
77	113·161 92		
78	115·054 01		
79	116·951 64		
80	118·854 73		

We wish to choose names 3 at a time: $r = 3$ and $(n - r) = 5 - 3 = 2$

$$_nC_r = \frac{n!}{r!(n - r)!} = \frac{5!}{3!2!} = \frac{120}{6 \times 2} = 10$$

10 different triplets of names can be found.

Partitions. There are $\begin{pmatrix} n \\ r_1, r_2, r_3, \ldots, r_k \end{pmatrix}$ ways of partitioning n objects into k groups of sizes $r_1, r_2, r_3, \ldots, r_k$, where $r_1 + r_2 + r_3 + \ldots + r_k = n$

and

$$\begin{pmatrix} n \\ r_1, r_2, r_3, \ldots, r_k \end{pmatrix} = \frac{n!}{r_1!r_2!r_3! \ldots r_n!} \tag{1.4.4}$$

▶ Example. We wish to send a group of 12 children to the zoo in 3 parties of size 5, 4 and 3. How many different arrangements are possible?

$$n = 12, \qquad r_1 = 5, \qquad r_2 = 4, \qquad r_3 = 3$$

check: $r_1 + r_2 + r_3 = 5 + 4 + 3 = 12 = n$

$$\binom{n}{r_1, r_2, r_3} = \binom{12}{5, 4, 3} = \frac{12!}{5!4!3!}$$

12! is a large number so log factorials will be used from table 1.4:

$$\log(12!) = 8 \cdot 6803, \qquad \log(5!) = 2 \cdot 0792, \qquad \log(4!) = 1 \cdot 3802,$$

$$\log(3!) = 0 \cdot 7782$$

$$\log\binom{n}{r_1, r_2, r_k} = \log(12!) - \log(5!) - \log(4!) - \log(3!)$$

$$= 8 \cdot 6803 - 2 \cdot 0792 - 1 \cdot 3802 - 0 \cdot 7782$$

$$= 4 \cdot 4427$$

$$\text{antilog}(4 \cdot 4427) = 10^4 \times 2 \cdot 77 = 27\,700 \quad \text{approx.}$$

There are approximately 27 700 ways of arranging 12 children into groups of 5, 4 and 3.

1.4.3 Empirical probability

Very often it is not possible to calculate the *a priori* probabilities associated with certain types of event. In these circumstances, we must carry out an experiment. If we have n trials and m of these produce the kind of event we are interested in we may compute

$$p = \frac{m}{n}$$

which is the *empirical probability* that these events will occur. The empirical probability is only an estimate of the *a priori* probability. The estimate is likely to get better and better as the value of n increases.

Addition Law. The probability that *either* of *two mutually exclusive* events will occur is equal to the sum of the probabilities of these events.

Thus, if $P_{(a)}$ is the probability that event a will occur, and $P_{(b)}$ is the probability that event b will occur, and $P_{(a\,or\,b)}$ is the probability that *either* event a or b will occur, then

$$P_{(a\,or\,b)} = P_{(a)} + P_{(b)} \tag{1.4.5}$$

Multiplication Law. The probability that *both* of *two independent* events will occur is equal to the product of the probabilities of these two events.

If $P_{(a\,and\,b)}$ is the probability that both events a and b will occur, then

$$P_{(a\,and\,b)} = P_{(a)} \cdot P_{(b)} \tag{1.4.6}$$

▶ Example. (i) At a conference the probability that a randomly chosen delegate will be (a) English is 0·4 and (b) American is 0·2. What is the probability that he will be either English or American?

$$P_{(a)} = 0 \cdot 4$$
$$P_{(a\,or\,b)} = 0 \cdot 4 + 0 \cdot 2 = 0 \cdot 6$$
$$P_{(b)} = 0 \cdot 2$$

(ii) If the probability that he is a young man is 0·4, what is the probability that he is a young American?

$$P_{(a)} = 0 \cdot 2$$
$$P_{(a\,and\,b)} = 0 \cdot 2 \times 0 \cdot 4 = 0 \cdot 08$$
$$P_{(b)} = 0 \cdot 4$$

2. Basic statistical procedures and frequency distributions

2.1 List of symbols

Roman letters

d	difference score
D	difference score; (sometimes) *largest* difference score (e.g., 4.4)
df	degrees of freedom
e	$2 \cdot 71828 \ldots$
E	expected frequency (e.g., 4.3)
f	(sometimes) degrees of freedom
F	statistic used in analysis of variance (7.0)
g_1	skew (2.4.3)
g_2	kurtosis (2.4.3)
H	statistic used in Friedman test (6.8)
H_0	null hypothesis (that the test assumptions are all valid) see Siegel (1956)
H_1	first hypothesis (alternative to H_0)
k	(often) number of groups
K	statistic used in K–S tests (e.g., 4.5) and Kruskal–Wallis test (5.8)
L	statistic used in Page's test (6.9)
M	mean of a sample (2.5)
M.S.	mean square (7.2.1)
n	group or sample size
n_t	number of scores contributing to a total (7.2.1)
N	(often) population size or total sample size (e.g., 7.2.1)
$N(\mu, \sigma)$	normal distribution with mean μ and variance σ^2 (2.7.2)
O	observed frequency (e.g., 4.3)
Q	statistic used in Cochran's test (6.3)
r	Pearson's product moment correlation coefficient (8.1)
S	standard deviation of a sample (2.5)
S^2	variance of a sample (2.5)
S.S.	sum of squares (7.2.1)
T	total
t	statistic used in t tests (e.g., 5.6) or subtotal
var.	variance
var. est.	variance estimate (2.5)
Z	standard scores (2.6)
z	unit normal deviate (2.6)

Greek letters

α	significance level (probability of type I error) (3.7.1)
β	probability of a type II error (3.7.1)
χ^2	statistic used in chi-square tests (e.g., 5.7)
μ	population mean (2.4.1)
ν	degrees of freedom
π	$3 \cdot 1416$ approx.
Π	product of ...
ρ	Spearman's measure of correlation (see Siegel, 1956)
Σ	summation of ... (2.2)
σ	population standard deviation (2.4.2)
σ^2	population variance (2.4.2)
τ	Kendall's measure of correlation (5.9)

Signs

$>$; $<$	greater than; less than
$\geq$; $\leq$	greater than or equal to; less than or equal to
$\neq$	not equal to
$\simeq$	approximately equal to
$\lvert a \rvert$	modulus of a, i.e., ignore sign and treat a as positive
$\hat{.\,.}$	estimate of ..., e.g., $\hat{\sigma}^2$ is variance estimate
$\overline{.\,.}$	mean of ..., e.g., $\bar{x}$ is mean of X scores
$\binom{n}{r}$ or $_nC_r$	number of combinations of r objects which can be chosen from n objects (1.4)
$\binom{n}{r_1, r_2, \ldots, r_j}$	number of partitions of n objects into groups of size $r_1, r_2, \ldots r_j$ (1.4)
$!$	factorial (1.4)

2.2   Subscript notation

When a letter (e.g., X) can refer to any one of a collection of scores, we use subscripts to identify which score is referred to at any particular time. A subscript is a letter, number or expression written *below* and to the *right* of the referenced symbol, e.g.,

$$X_1, X_2, X_i, X_{(n-1)}, X_{(i+1)}, \quad \text{etc.}$$

Their use in algebra is straightforward, but practice is required before they can be used with ease.

Single subscripts are used when the collection of scores can be set out in a single array thus:

Number:	1	2	3	4	5	6	...	$n-2$	$n-1$	n
collection of scores:	47	6	48	2	39	90	...	17	50	69
X:	X_1	X_2	X_3	X_4	X_5	X_6	...	$X_{(n-2)}$	$X_{(n-1)}$	X_n

These are n scores and X can refer to any one of them. The subscript specifies clearly which one is required. The subscript n, however, is undefined until we know how large n is; similarly X_i is undefined until i is specified. The summation symbol is a device for specifying the subscript as it changes:

$$\sum_{i=1}^{n} X_i = X_1 + X_2 + X_3 + X_4 + \ldots + X_{(n-2)} + X_{(n-1)} + X_n$$

Here, the value of i begins at 1 and increases by steps of 1 until it reaches n, that is it takes on successively the values $1, 2, 3, 4, \ldots, (n-1), n$.

This is a convenient device which allows us express briefly but unambiguously such mathematical operations as the computation of the mean and variance

$$\mu = \frac{\sum_{i=1}^{n} X_i}{n}$$

$$\sigma^2 = \frac{\sum_{i=1}^{n} (X_i - \mu)^2}{n}$$

Often, such precision is unnecessary and the subscript paraphernalia is dropped without causing ambiguity:

$$\mu = \frac{\Sigma X}{n} \qquad \sigma^2 = \frac{\Sigma (X - \mu)^2}{n}$$

The great value of subscript notation becomes evident when the values of X are arranged in two dimensional arrays. Here it is convenient to indicate a particular value of X by specifying the row and column it belongs to. The convention is that *rows are specified first and columns second.*

$$X_{3,7}$$

belongs to row 3 and column 7.

When the rows and columns are undefined, they are commonly indicated by i and j respectively:

$$X_{ij}$$

The various operations will be illustrated with reference to this example

		Columns		
	1	2	3	Totals
Rows 1	16	71	14	101
2	98	61	35	194
3	29	81	15	125
4	72	10	11	93
Totals	215	223	75	513

It can be seen easily that

$$X_{1,1} = 16, \qquad X_{1,2} = 71, \qquad X_{2,1} = 98$$

The expression

$$\sum_{i=1}^{4} X_{i,2}$$

refers to the total of scores in the second column. The summation asks us to change the value of i from 1 to 2 to 3 to 4 while summing. This leaves the value of the column number the same (2). Thus:

$$\sum_{i=1}^{4} X_{i,2} = X_{1,2} + X_{2,2} + X_{3,2} + X_{4,2} = 223$$

Similarly,

$$\sum_{j=1}^{3} X_{1,j} = 101$$

because it refers to the first row total. Notice that the row number stays the same (1) while the column number varies. We can see that

$$\sum_{i=1}^{m} X_{ij}$$

refers to a column total but we do not know which column until j is specified. We call it the jth column total. Similarly,

$$\sum_{j=1}^{n} X_{ij}$$

is the ith row total.

The grand total can be expressed as the sum of the column totals

$$T = \sum_{j=1}^{n} \sum_{i=1}^{m} X_{ij}$$

where there are m rows and n columns, or as sum of the row totals

$$T = \sum_{i=1}^{m} \sum_{j=1}^{n} X_{ij}$$

Clearly, a sharp eye must be kept on the 'small print' when dealing with subscripts.

Dot notation. We define the row mean above as

$$i\text{th row mean} = \frac{\sum_{j=1}^{n} X_{ij}}{n}$$

This is sometimes written as $\bar{X}_{i\cdot}$. The dot, which replaces the j, indicates that the summation involved all the scores in the ith row (i.e., all values of j were used but the value of i was held steady). Similarly,

$$\bar{X}_{\cdot j} = \frac{\sum\limits_{i=1}^{m} X_{ij}}{m} \quad \text{is the } j\text{th column average.}$$

The bar above $\bar{X}_{\cdot j}$ indicates that it is an average (mean) value, rather than a total.

We can extend this a little further:

$$\bar{X}_{\cdot\cdot} = \frac{\sum\limits_{i=1}^{m}\sum\limits_{j=1}^{n} X_{ij}}{mn} = \frac{\sum\limits_{j=1}^{n}\sum\limits_{i=1}^{m} X_{ij}}{mn}$$

which is the grand mean.

Dot notation is elegant but often intellectually taxing, while subscript notation is cumbersome and taxing. Both are very valuable to statisticians when developing techniques but are largely unnecessary for summarizing the bulk of the elementary statistical material. Such notation has been dropped whenever possible in this book.

20

2.3 Ranking procedures

In this book, we follow the convention that the lowest score in a group gets rank 1.

Score	17	18	23	24	79	80	81	Sum of ranks
Rank	1	2	3	4	5	6	7	28

Tied scores share the ranks which would have been given them if the scores had been slightly different:

Score	17	17	23	23	80	80	80	Sum of ranks
Rank	1·5	1·5	3·5	3·5	6	6	6	28

These ranks can now be treated as if they were ordinary (untied) ranks. Notice that the sum of the ranks is the same with and without ties. The variance of ranks is, however, always less when ties are present. Many statistical tests assume that there are no ties and corrections need to be applied to tied data. The effect of these corrections is very small in most cases and they have been omitted from the book as unnecessary complications. Failure to correct for ties usually makes the results seem slightly less significant. In this way, any error is on the conservative side.

As a check on ranking we can calculate the sum of N ranks as

$$\frac{N(N + 1)}{2} \tag{2.3.1}$$

For example, in the above case we have $N = 7$. As a result, we expect the sum of ranks to be $7 \times 8/2 = 28$, which it is.

The mean (M) of a set of N ranks is:

$$M = \frac{N + 1}{2} \tag{2.3.2}$$

The variance (var.) of a set of N ranks is:

$$\text{var.} = \frac{N^2 - 1}{12} \tag{2.3.3}$$

2.4   Population descriptions

The statistics in sections 2.4 and 2.5 are for describing collections of scores.

When a collection of scores is complete we call it a *population*.

Descriptions of various aspects of a population (central tendency, dispersion, etc.) are called *parameters*.

When a collection of scores is incomplete we call it a *sample*.

When samples are taken at random from a population, we may use the sample scores to produce *estimates* of population parameters.

None of the formulae in sections 2.4 to 2.6 make any assumptions about the distribution characteristics of the population. In particular, it should be noted that these formulae hold when the scores are *not normally distributed*.

Unless otherwise specified, the symbol X refers to the individual scores which make up the population or sample which contains N such scores and:

ΣX is the sum of all N scores
ΣX^2 is the sum *of the squares* of all N scores
$(\Sigma X)^2$ is the *square* of the sum of all N scores

2.4.1   Central value

Arithmetic mean (μ).

$$\mu = \frac{\Sigma X}{N} \tag{2.4.1a}$$

The mean has two special properties:
1. The sum of the deviations of each score from the mean is zero

$$\Sigma(X - \mu) = 0 \tag{2·4.1b}$$

2. The sum of the squares of the deviations of each score from the mean is a minimum

$$\Sigma(X - \mu)^2 = \text{Minimum} \tag{2.4.1c}$$

These properties qualify the mean as the most useful measure of central value.

Median.   The median is that point on a scale which is exceeded by only 50 per cent of the scores in the population. In a collection of N scores the median is the value associated with the score which has rank $(N + 1)/2$. When N is even, the median is the point midway between the scores which have ranks $N/2$ and $(N + 2)/2$.

Mode.   The *crude mode* is the most frequently occurring category of score.

The *computed mode* which is not necessarily the same as the crude mode may be found:

$$\text{Mode} = 3 \times \text{Median} - 2 \times \text{Mean} \tag{2.4.1d}$$

The mode can only be computed in this way if the scores are at least rankable.

Choosing a measure of central tendency.   The scale in use restricts the choice.

Continuous scale: all three acceptable
Rank scale: median and mode only
Nominal scale: mode only.

The general order of preference is mean > median > mode, when the distribution is symmetrical. For non-symmetrical distributions the median is often the most appropriate statistic.

▶ Example. A test of abstract reasoning was given to a class of 30 students. Their performance was given a percentage mark. Give the mean, median and modes of their scores.

Score	Rank
13	1
21	2
24	3
27	4
27	5
29	6
30	7
31	8
31	9
33	10
34	11
37	12
38	13
39	14
41	15
44	16
44	17
46	18
47	19
53	20
53	21
58	22
61	23
64	24
67	25
70	26
77	27
78	28
81	29
83	30

Total 1381

$N = 30$

(a) $Mean = \dfrac{\Sigma X}{N} = \dfrac{1381}{30} = 46 \cdot 0$

(b) Median: N is even; find scores with rank $N/2$, $(N + 2)/2$

$N/2 = 30/2 = 15$: score is 41

$(N + 2)/2 = 32/2 = 16$: score is 44

$Median = (41 + 44)/2 = 42 \cdot 5$

(c) Mode: The scores have been broken into intervals of width 10. The frequencies are:

10–19: 1
20–29: 5
30–39: 8
40–49: 5
50–59: 3
60–69: 3
70–79: 3
80–89: 2

The most common interval is 30–39. We put the mode at the midpoint of this interval:

$Crude\ mode = 34 \cdot 5$

The computed mode is

$Mode = 3\ Median - 2\ Mean$

$= 3 \times 42 \cdot 5 - 2 \times 46 \cdot 0$

$Computed\ mode = 35 \cdot 5$

2.4.2 Dispersion

Range Where X_1 is the smallest score and X_N is the largest score

$$\text{Range} = X_N - X_1 \qquad (2.4.2a)$$

Interquartile range and quartile deviation. Where Q_1 is the cut-off point for the lower 25 per cent of scores and Q_3 is the cut-off point for the top 25 per cent of scores:

$$\text{Interquartile range} = Q_3 - Q_1 \qquad (2.4.2b)$$

$$\text{Quartile deviation} = \frac{Q_3 - Q_1}{2} \qquad (2.4.2c)$$

Variance: var, σ^2. Where each score (X) is expressed as a deviation (D) from the mean (μ)

$$D = X - \mu$$

the variance is:

$$\sigma^2 = \frac{\Sigma D^2}{N} = \frac{\Sigma(X - \mu)^2}{N} \qquad (2.5.2d)$$

An alternative computational formula is

$$\sigma^2 = \frac{\Sigma X^2}{N} - \left(\frac{\Sigma X}{N}\right)^2 \qquad (2.4.2e)$$

This formula does not require that the scores be converted to deviation scores. The variance is given in *square units*.

Standard deviation: σ. The standard deviation is the square root of the variance

$$\sigma = \sqrt{\text{var.}} = \sqrt{\left[\frac{\Sigma X^2}{N} - \left(\frac{\Sigma X}{N}\right)^2\right]}$$

It is this relationship which makes a special symbol of variance unnecessary since

$$\text{var.} = \sigma^2$$

Choosing a measure of dispersion. The range is used *only* when interest is focused on the extreme points of the scale. The interquartile range has the advantage of being easily computed and is of value in conjunction with the median when it is necessary to divide the population into equal sized groups.

The variance is an attractive measure from a mathematical point of view and is used for preference in nearly all advanced statistical exercises. The variance is, however, expressed in terms of square units and the standard deviation (which uses simple units) is a more intuitively acceptable description.

▶ Example. A test of abstract reasoning was given to a class of 30 students. Give the range, interquartile range, quartile deviation, variance and standard deviation of their scores.

Score	Rank	
13	1	*Range* $= 83 - 13 = 70$
21	2	*Interquartile range:* $N = 30$, 25 per cent of the scores involve $30/4 = 7.5$ scores.
24	3	Scores 30 and 31 have ranks 7 and 8.
27	4	
27	5	Q_1 lies midway between 30 and 31 $= 30.5$.
29	6	Scores 64 and 61 have ranks 24 and 23.
30	7	Q_3 lies midway between 64 and 61 $= 62.5$.

Q_1 ——————

Score	Rank	
		Interquartile range $= 62.5 - 30.5 = 32$
31	8	
31	9	*Quartile deviation* $= 32/2 = 16$
33	10	
34	11	Variance $= \dfrac{\Sigma X^2}{N} - \left(\dfrac{\Sigma X}{N}\right)^2$
37	12	
38	13	$= \dfrac{74\,415}{30} - \left(\dfrac{1381}{30}\right)^2$
39	14	
41	15	$= 24\,805 - 2119.1$
44	16	
44	17	*Variance* $= 361.4$
46	18	
47	19	Standard deviation, $\sigma = \sqrt{\text{Variance}} = \sqrt{361.4} = 19$
53	20	
53	21	
58	22	
61	23	

Q_2 ——————

Score	Rank
64	24
67	25
70	26
77	27
78	28
81	29
83	30

$\Sigma X = 1381$

$\Sigma X^2 = 74\,415$

2.4.3 Other measures of distribution

Skew: g_1.   Skew is a measure of asymmetry. Where $D = X - \mu$

$$g_1 = \frac{\Sigma D^3}{\Sigma D^2}\sqrt{\frac{N}{\Sigma D^2}} \qquad (2.4.3a)$$

When the majority of scores lie *below* the mean the skew is positive and $g_1 > 0$.

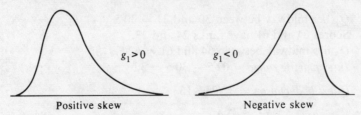

$g_1 > 0$ $g_1 < 0$

Positive skew   Negative skew

Fig. 2.4.3a   Examples of distributions with positive and negative skew.

Kurtosis: g_2.   Kurtosis describes whether the distribution is flat or peaked. Where $D = X - \mu$

$$g_2 = \frac{N\Sigma D^4}{[\Sigma D^2]^2} - 3 \qquad (2.4.3b)$$

Then the distribution has a pronounced peak (leptokurtic) $g_2 > 0$.

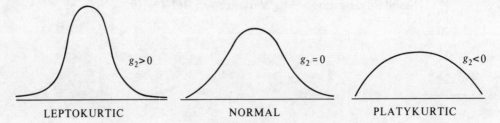

$g_2 > 0$ $g_2 = 0$ $g_2 < 0$

LEPTOKURTIC   NORMAL   PLATYKURTIC

Fig. 2.4.3b   Examples of distributions with differing degrees of kurtosis.

▶ Example.   A test of abstract reasoning was given to a class of 30 students. Their performance was given a percentage mark. Give the skew and kurtosis of the resulting distribution.

Score	D
13	−33
21	−25
24	−22
27	−19
27	−19
29	−17
30	−16
31	−15
31	−15
33	−13
34	−12
37	−9
38	−8
39	−7
41	−5
44	−2
44	−2
46	0
47	1
53	7
53	7
58	12
61	15
64	18
67	21
70	24
77	31
78	32
81	35
83	37

$N = 30$

Mean $= 46$

$\Sigma D^2 = 10\,843$

$\Sigma D^3 = 93\,457$

$\Sigma D^4 = 8\,373\,507$

Skew

$$g_1 = \frac{\Sigma D^3}{\Sigma D^2}\sqrt{\frac{N}{\Sigma D^2}}$$

$$= \frac{93\,457}{10\,843}\sqrt{\frac{30}{10\,843}} = 0.45$$

Kurtosis

$$g_2 = \frac{N\Sigma D^4}{[\Sigma D^2]^2} - 3$$

$$= \frac{30 \times 8\,373\,507}{10\,843^2} - 3$$

$$= -0.863$$

2.5 Estimating from samples

Samples are generally taken with the express purpose of estimating the parameters of a population, in particular the mean and variance. Consequently, we are rarely interested in the mean and variance of the sample itself, only the *estimates* of the population values.

   We use the following notation to distinguish among true population values (often unknown), estimates based on samples and sample values.

Value	True population value	Estimate from sample	Sample value
Mean	μ	$\hat{\mu}$	$\overline{X}$
Variance	σ^2	$\hat{\sigma}^2$	S^2

$\hat{\mu}$ is read 'mew cap', and $\hat{\sigma}^2$ is read 'sigma squared cap'.

The sample size is n while the population size is N. Samples are here assumed to be *random* samples.

2.5.1 Estimate of the population mean

The sample mean is, in fact, the best estimate of the population mean:

$$\hat{\mu} = \frac{\Sigma X}{n} = \overline{X} \tag{2.5.1}$$

2.5.2 Estimate of the population variance

Calculating the variance requires a knowledge of the mean. In these circumstances, where *the true population mean is known:*

$$\hat{\sigma}^2 = \frac{\Sigma(X - \mu)^2}{n} \tag{2.5.2a}$$

Usually, it is *not* known and we must use the best estimate of the mean ($\hat{\mu}$)

$$\hat{\sigma}^2 = \frac{\Sigma(X - \hat{\mu})^2}{n - 1} \tag{2.5.2b}$$

Using only an *estimate* of the population mean incurs the loss of 1 degree of freedom. This is why we divide by $n - 1$ rather than n.

   A convenient computational formula is:

$$\hat{\sigma}^2 = \frac{\Sigma X^2}{n - 1} - \frac{(\Sigma X)^2}{n(n - 1)} \tag{2.5.2c}$$

$$= \frac{\Sigma X^2 - (\Sigma X)^2/n}{n - 1} \tag{2.5.2d}$$

▶ Example. Treat the scores in example 2.4.2 as a sample and estimate the population mean, variance and standard deviation.

$$\Sigma X = 1381; \qquad \Sigma X^2 = 74\,415; \qquad n = 30$$

Population mean ($\hat{\mu}$)

$$\hat{\mu} = \frac{\Sigma X}{n} = \frac{1381}{30} = 46.0$$

Population variance ($\hat{\sigma}^2$)

$$\hat{\sigma}^2 = \frac{\Sigma X^2 - (\Sigma X)^2/n}{n-1} = \frac{74\,415 - 1381^2/30}{29} = 373.9$$

Standard deviation

$$\hat{\sigma} = \sqrt{\hat{\sigma}^2} = \sqrt{373.9} = 19.3$$

2.5.3 Standard error of the sample mean

When the true mean and variance of a population is known (μ, σ^2) we may make some inferences about the results of *repeated* sampling from that population. Where $\bar{x}$ is the mean of a single sample, then over *many* such samples we expect its average value to be:

$$\mu_{\bar{x}} = \mu \qquad (2.5.3\text{a})$$

The sample value, $\bar{x}$, will vary about the population mean μ with standard error:

$$\sigma_{\bar{x}} = \frac{\sigma}{\sqrt{n}} = \sqrt{\frac{\sigma^2}{n}} \qquad (2.5.3\text{b})$$

where n is the sample size.

▶ Example.   If repeated random samples of size 20 are taken from a population with mean 46·0 and variance 374, what will be the mean and standard error of the sample means?

$$\mu = 46\cdot0; \qquad \sigma^2 = 374; \qquad n = 20$$

(a) Mean of sample means

$$\mu_{\bar{x}} = \mu = 46\cdot0$$

(b) Standard error of the sample means

$$\sigma_{\bar{x}} = \sqrt{\frac{\sigma^2}{n}} = \sqrt{\frac{374}{20}} = 4\cdot3$$

2.6   Standard scores

A score X may be expressed relative to other scores in its population as a *standard score* (Z):

$$Z = \frac{X - \mu}{\sigma} \qquad (2.6.1)$$

where μ is the mean of the population and σ is its standard deviation.

The upper case letter (Z) is used in this book to indicate standard scores. Unit normal deviates (see section 2.7.2) are a special type of standard score and are denoted by the lower case letter (z).

Properties of standard scores.   For a population containing N standard scores:

$$\Sigma Z = 0 \quad \text{and} \quad \Sigma Z^2 = N$$

It follows that the mean and variance are:

$$\mu_z = \frac{\Sigma Z}{N} = 0$$

$$\sigma_z^2 = \frac{\Sigma Z^2}{N} - \left(\frac{\Sigma Z}{N}\right)^2 = \frac{N}{N} = 1$$

Thus, a population of standard scores has zero mean ($\mu = 0$) and unit variance ($\sigma^2 = 1$).

▶ Example.   Thirty students are given an abstract reasoning test. Their mean score is 46·0 with a variance of 361·4. One student has a score of 13. Express this as a standard score.

$$\mu = 46\cdot0, \qquad \sigma^2 = 361\cdot4, \qquad \sigma = \sqrt{361\cdot4} = 19, \qquad X = 13$$

$$Z = \frac{X - \mu}{\sigma} = \frac{13 - 46}{19} = -1\cdot74$$

2.7 Standard frequency distributions

2.7.1 Frequency distributions

Table 2.7.1a is a *frequency table* which describes the theoretical expectation of the outcomes of 100 experiments where a coin is tossed five times. The outcome of each experiment is expressed in terms of the number of coins which land showing heads.

Table 2.7.1   Frequency table (see text).

Number of heads in 5 tosses	0	1	2	3	4	5	Total
Frequency	3	16	31	31	16	3	100

Figure 2.7.1 is its frequency histogram. Each column corresponds to a possible outcome in the coin tossing experiment. The *area* of each column corresponds to the *frequency* of its corresponding outcome. The area is expressed as a *percentage* of the total area. We may *combine areas* to describe the frequency of alternative outcomes. Thus, the percentage of experiments which yield 4 or more heads can be found by combining the area of two right hand columns (16 + 3 per cent = 19 per cent).

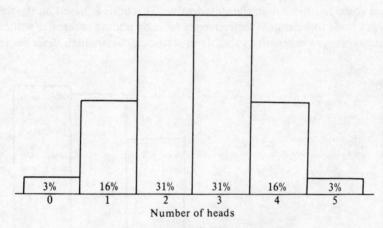

Fig. 2.7.1   Frequency histogram (see table 2.1).

As the number of possible outcomes is increased the columns become narrower and the tops of the columns nearer to their neighbours. It is convenient then to replace the frequency histogram by a smooth curve. Figure 2.7.1b gives the frequency distribution of possible outcomes of an experiment where 100 coins are tossed. The shaded area represents the percentage of outcomes we expect to yield *more* than 60 heads.

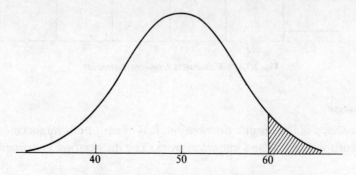

Fig. 2.7.1b   Frequency distribution (see text).

The smooth curve illustrated in Fig. 2.7.1b is only an *approximation* to the frequency histogram. As the number of possible results increases the approximation gets better. While some accuracy is lost by assuming the curve to be continuous and smooth, the loss is often very small and our assumption can result in a considerable saving in computation.

The non-parametric statistics to be discussed in this book have frequency histograms which rapidly approximate well known distributions as the experimental samples increase in size. For *small samples*, we base our inferences on *exact* frequency histograms. For *large samples*, we base our inferences on *approximations to smooth frequency distributions* such as normal, F, chi square, Poisson and Student's t distributions. Tables for these distributions are given in sections 2.7.2 to 2.7.7.

Statistical inference is commonly based on the frequency of extreme scores. It can be seen from Fig. 2.7.1b that it is rare to observe *more than* 60 heads when throwing 100 coins. The area which represents the frequency of obtaining a result *as extreme or more extreme* is called *the tail of the distribution*. If we are interested in extreme results on *both* sides of the mean we need the area in *both tails*. It is this terminology which gives rise to the expressions 1-TAIL and 2-TAIL tests.

Table 2.7.1b Cumulative frequency table

Proportion of heads	$\frac{0}{5}$	$\frac{1}{5}$	$\frac{2}{5}$	$\frac{3}{5}$	$\frac{4}{5}$	$\frac{5}{5}$
Frequency	3	19	50	81	97	100

Table 2.7.1b is a *cumulative frequency* distribution. It states the frequency with which our variable (number of heads) takes values equal to or less than the stated value. This table is based on the figures in table 2.7.1. We convert a frequency table to a cumulative frequency table by adding successive values from left to right. Cumulative frequency tables are especially valuable in statistical tests which describe and compare distributions.

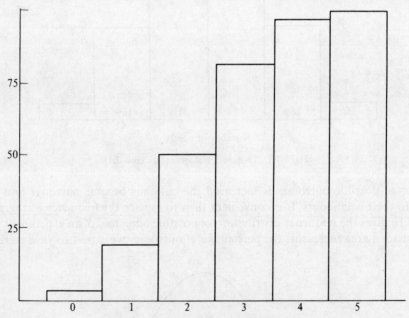

Fig. 2.7.1c Cumulative frequency histogram.

2.7.2 Normal distribution

Figure 2.7.2 shows the shape of the normal distribution. It is of enormous importance in statistics because it occurs so often in both theoretical and empirical work. The distribution is defined by its mean (μ) and standard deviation (σ).

The symbol $N(\mu, \sigma)$ is reserved for a distribution which is normal (N), has a mean, μ, and standard deviation, σ. The distribution described by table 2.7.2 is $N(0, 1)$—normally distributed with a mean of zero and standard deviation equal to $1 \cdot 00$.

In this book, *standard scores* which are normally distributed are denoted by a *lower case letter z*. These are often called *unit normal deviates* because (a) they have a standard deviation of one (unity), (b) they are normally distributed and (c) they are deviations from the population mean.

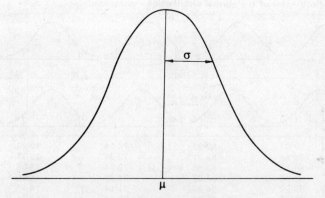

Fig. 2.7.2　Normal distribution, $N(0, 1)$.

The height of the normal curve against values of z is given in the end of column of table 2.7.2. The *area under the normal curve* between two points is found in table 2.7.2 which gives percentage areas of various kinds as indicated by the diagrams at the top of the columns. The cut-off line is defined by the value z which is given in column 1.

Many of the important properties of the normal distribution can be derived from the *central limit theorem* which says that the possible sums of a set of independently chosen scores will be normally distributed when the set of scores is large. Since many effects in nature are a compound result of many separate influences, it follows that measures of these effects are often normally distributed.

Table 2.7.2　Areas and ordinates of the normal distribution.

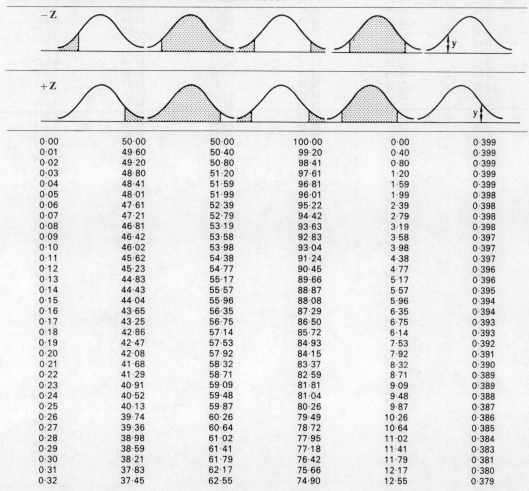

					y
0·00	50·00	50·00	100·00	0·00	0·399
0·01	49·60	50·40	99·20	0·40	0·399
0·02	49·20	50·80	98·41	0·80	0·399
0·03	48·80	51·20	97·61	1·20	0·399
0·04	48·41	51·59	96·81	1·59	0·399
0·05	48·01	51·99	96·01	1·99	0·398
0·06	47·61	52·39	95·22	2·39	0·398
0·07	47·21	52·79	94·42	2·79	0·398
0·08	46·81	53·19	93·63	3·19	0·398
0·09	46·42	53·58	92·83	3·58	0·397
0·10	46·02	53·98	93·04	3·98	0·397
0·11	45·62	54·38	91·24	4·38	0·397
0·12	45·23	54·77	90·45	4·77	0·396
0·13	44·83	55·17	89·66	5·17	0·396
0·14	44·43	55·57	88·87	5·57	0·395
0·15	44·04	55·96	88·08	5·96	0·394
0·16	43·65	56·35	87·29	6·35	0·394
0·17	43·25	56·75	86·50	6·75	0·393
0·18	42·86	57·14	85·72	6·14	0·393
0·19	42·47	57·53	84·93	7·53	0·392
0·20	42·08	57·92	84·15	7·92	0·391
0·21	41·68	58·32	83·37	8·32	0·390
0·22	41·29	58·71	82·59	8·71	0·389
0·23	40·91	59·09	81·81	9·09	0·389
0·24	40·52	59·48	81·04	9·48	0·388
0·25	40·13	59·87	80·26	9·87	0·387
0·26	39·74	60·26	79·49	10·26	0·386
0·27	39·36	60·64	78·72	10·64	0·385
0·28	38·98	61·02	77·95	11·02	0·384
0·29	38·59	61·41	77·18	11·41	0·383
0·30	38·21	61·79	76·42	11·79	0·381
0·31	37·83	62·17	75·66	12·17	0·380
0·32	37·45	62·55	74·90	12·55	0·379

Table 2.7.2 Areas and ordinates of the normal distribution—continued.

−Z

+Z

Z					y
0·33	37·07	62·93	74·14	12·93	0·378
0·34	36·69	63·31	73·39	13·31	0·377
0·35	36·32	63·68	72·64	13·68	0·375
0·36	35·94	64·06	71·89	14·06	0·374
0·37	35·57	64·43	71·14	14·43	0·373
0·38	35·20	64·80	70·40	14·80	0·371
0·39	34·83	65·16	69·66	15·17	0·370
0·40	34·46	65·54	68·92	15·54	0·368
0·41	34·09	65·91	68·18	15·91	0·367
0·42	33·73	66·27	67·45	16·27	0·365
0·43	33·36	66·64	66·72	16·64	0·364
0·44	33·00	67·00	66·00	17·00	0·362
0·45	32·64	67·36	65·27	17·36	0·361
0·46	32·28	67·72	64·55	17·72	0·359
0·47	31·92	68·08	63·84	18·08	0·357
0·48	31·56	68·44	63·13	18·44	0·356
0·49	31·21	68·79	63·42	18·79	0·354
0·50	30·85	69·15	61·71	19·15	0·352
0·51	30·50	69·50	61·01	19·50	0·350
0·52	30·15	69·85	60·31	19·85	0·348
0·53	29·81	70·19	59·61	20·19	0·347
0·54	29·46	70·54	58·92	20·54	0·345
0·55	29·12	70·88	58·23	20·88	0·343
0·56	28·78	71·22	57·55	21·22	0·341
0·57	28·44	71·56	56·78	21·56	0·339
0·58	28·10	71·90	56·19	21·90	0·337
0·59	27·76	72·24	55·52	22·24	0·335
0·60	27·43	72·57	54·85	22·57	0·333
0·61	27·09	72·91	54·19	22·91	0·331
0·62	26·76	73·24	53·53	23·24	0·329
0·63	26·44	73·56	52·87	23·56	0·327
0·64	26·11	73·89	52·22	23·89	0·325
0·65	25·79	74·21	51·57	24·21	0·323
0·66	25·46	74·54	50·93	24·54	0·321
0·67	25·14	74·86	50·29	24·86	0·319
0·68	24·83	75·17	49·65	25·17	0·317
0·69	24·51	75·49	49·02	25·49	0·314
0·70	24·20	75·80	48·40	25·80	0·312
0·71	23·89	76·11	47·77	26·11	0·310
0·72	23·58	76·42	47·15	26·42	0·308
0·73	23·27	76·73	46·54	26·73	0·306
0·74	22·97	77·03	45·93	27·03	0·303
0·75	22·66	77·34	45·33	27·34	0·301
0·76	22·36	77·64	44·73	27·64	0·299
0·77	22·07	77·93	44·13	27·93	0·297
0·78	21·77	78·23	54·54	28·23	0·294
0·79	21·48	78·52	42·96	28·52	0·292
0·80	21·19	78·81	42·37	28·81	0·290
0·81	20·90	79·10	41·80	29·10	0·287
0·82	20·61	79·39	41·22	29·39	0·285
0·83	20·33	79·67	40·66	29·67	0·283
0·84	20·05	79·95	40·09	29·95	0·280
0·85	19·77	80·23	39·53	30·23	0·278
0·86	19·49	80·51	38·98	30·51	0·276
0·87	19·22	80·78	38·43	30·78	0·273
0·88	18·94	81·06	37·89	31·06	0·271
0·89	18·67	81·33	37·35	31·33	0·268
0·90	18·41	81·59	36·81	31·59	0·266
0·91	18·14	81·86	36·28	31·86	0·264
0·92	17·88	82·18	35·76	32·12	0·261
0·93	17·62	82·38	35·24	32·38	0·259
0·94	17·36	83·64	34·72	32·64	0·256
0·95	17·11	82·89	34·21	32·89	0·254
0·96	16·85	83·15	33·71	33·15	0·252
0·97	16·60	83·40	33·21	33·40	0·249
0·98	16·36	83·64	32·71	33·64	0·247
0·99	16·11	83·89	32·22	33·89	0·244
1·00	15·87	84·13	31·73	34·13	0·242
1·01	15·63	84·37	31·25	34·37	0·240
1·02	15·39	84·61	30·78	34·61	0·237

Table 2.7.2—continued.

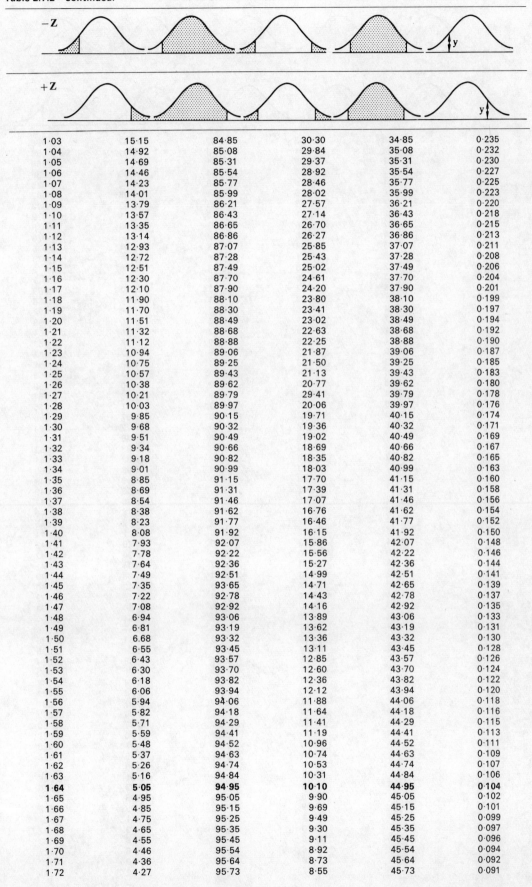

1·03	15·15	84·85	30·30	34·85	0·235
1·04	14·92	85·08	29·84	35·08	0·232
1·05	14·69	85·31	29·37	35·31	0·230
1·06	14·46	85·54	28·92	35·54	0·227
1·07	14·23	85·77	28·46	35·77	0·225
1·08	14·01	85·99	28·02	35·99	0·223
1·09	13·79	86·21	27·57	36·21	0·220
1·10	13·57	86·43	27·14	36·43	0·218
1·11	13·35	86·65	26·70	36·65	0·215
1·12	13·14	86·86	26·27	36·86	0·213
1·13	12·93	87·07	25·85	37·07	0·211
1·14	12·72	87·28	25·43	37·28	0·208
1·15	12·51	87·49	25·02	37·49	0·206
1·16	12·30	87·70	24·61	37·70	0·204
1·17	12·10	87·90	24·20	37·90	0·201
1·18	11·90	88·10	23·80	38·10	0·199
1·19	11·70	88·30	23·41	38·30	0·197
1·20	11·51	88·49	23·02	38·49	0·194
1·21	11·32	88·68	22·63	38·68	0·192
1·22	11·12	88·88	22·25	38·88	0·190
1·23	10·94	89·06	21·87	39·06	0·187
1·24	10·75	89·25	21·50	39·25	0·185
1·25	10·57	89·43	21·13	39·43	0·183
1·26	10·38	89·62	20·77	39·62	0·180
1·27	10·21	89·79	29·41	39·79	0·178
1·28	10·03	89·97	20·06	39·97	0·176
1·29	9·85	90·15	19·71	40·15	0·174
1·30	9·68	90·32	19·36	40·32	0·171
1·31	9·51	90·49	19·02	40·49	0·169
1·32	9·34	90·66	18·69	40·66	0·167
1·33	9·18	90·82	18·35	40·82	0·165
1·34	9·01	90·99	18·03	40·99	0·163
1·35	8·85	91·15	17·70	41·15	0·160
1·36	8·69	91·31	17·39	41·31	0·158
1·37	8·54	91·46	17·07	41·46	0·156
1·38	8·38	91·62	16·76	41·62	0·154
1·39	8·23	91·77	16·46	41·77	0·152
1·40	8·08	91·92	16·15	41·92	0·150
1·41	7·93	92·07	15·86	42·07	0·148
1·42	7·78	92·22	15·56	42·22	0·146
1·43	7·64	92·36	15·27	42·36	0·144
1·44	7·49	92·51	14·99	42·51	0·141
1·45	7·35	93·65	14·71	42·65	0·139
1·46	7·22	92·78	14·43	42·78	0·137
1·47	7·08	92·92	14·16	42·92	0·135
1·48	6·94	93·06	13·89	43·06	0·133
1·49	6·81	93·19	13·62	43·19	0·131
1·50	6·68	93·32	13·36	43·32	0·130
1·51	6·55	93·45	13·11	43·45	0·128
1·52	6·43	93·57	12·85	43·57	0·126
1·53	6·30	93·70	12·60	43·70	0·124
1·54	6·18	93·82	12·36	43·82	0·122
1·55	6·06	93·94	12·12	43·94	0·120
1·56	5·94	94·06	11·88	44·06	0·118
1·57	5·82	94·18	11·64	44·18	0·116
1·58	5·71	94·29	11·41	44·29	0·115
1·59	5·59	94·41	11·19	44·41	0·113
1·60	5·48	94·52	10·96	44·52	0·111
1·61	5·37	94·63	10·74	44·63	0·109
1·62	5·26	94·74	10·53	44·74	0·107
1·63	5·16	94·84	10·31	44·84	0·106
1·64	**5·05**	**94·95**	**10·10**	**44·95**	**0·104**
1·65	4·95	95·05	9·90	45·05	0·102
1·66	4·85	95·15	9·69	45·15	0·101
1·67	4·75	95·25	9·49	45·25	0·099
1·68	4·65	95·35	9·30	45·35	0·097
1·69	4·55	95·45	9·11	45·45	0·096
1·70	4·46	95·54	8·92	45·54	0·094
1·71	4·36	95·64	8·73	45·64	0·092
1·72	4·27	95·73	8·55	45·73	0·091

Table 2.7.2 Areas and ordinates of the normal distribution—continued.

1·73	4·18	95·82	8·37	45·82	0·089
1·74	4·09	95·91	8·19	45·91	0·088
1·75	4·01	95·99	8·01	45·99	0·086
1·76	3·92	96·08	7·84	46·08	0·085
1·77	3·84	96·16	7·68	46·16	0·083
1·78	3·75	96·25	7·51	46·25	0·082
1·79	3·67	96·33	7·35	46·33	0·080
1·80	3·59	96·41	7·19	46·41	0·079
1·81	3·52	96·48	7·03	46·48	0·078
1·82	3·44	96·56	6·88	46·56	0·076
1·83	3·36	96·64	6·73	46·64	0·075
1·84	3·29	96·71	6·58	46·71	0·073
1·85	3·22	96·78	6·43	46·78	0·072
1·86	3·15	96·85	6·29	46·85	0·071
1·87	3·08	96·92	6·15	46·92	0·069
1·88	3·01	96·99	6·01	46·99	0·068
1·89	2·94	97·06	5·88	47·06	0·067
1·90	2·87	97·13	5·75	47·13	0·066
1·91	2·81	97·19	5·62	47·19	0·064
1·92	2·74	97·26	5·49	47·26	0·063
1·93	2·68	97·32	5·36	47·32	0·062
1·94	2·62	97·38	5·24	47·38	0·061
1·95	2·56	97·44	5·12	47·44	0·060
1·96	**2·50**	**97·50**	**5·00**	**47·50**	**0·058**
1·97	2·44	97·56	4·89	47·56	0·057
1·98	2·39	97·61	4·77	47·61	0·056
1·99	2·33	97·67	4·66	47·67	0·055
2·00	2·28	97·72	4·55	47·72	0·054
2·01	2·22	97·78	4·45	47·78	0·053
2·02	2·17	97·83	4·34	47·83	0·052
2·03	2·12	97·88	4·24	47·88	0·051
2·04	2·07	97·93	4·14	47·93	0·050
2·05	2·02	97·98	4·04	47·98	0·049
2·06	1·97	98·03	3·94	48·03	0·048
2·07	1·92	98·08	3·85	48·08	0·047
2·08	1·88	98·12	3·75	48·12	0·046
2·09	1·83	98·17	3·66	48·17	0·045
2·10	1·79	98·21	3·58	48·21	0·044
2·11	1·74	98·26	3·49	48·26	0·043
2·12	1·70	98·30	3·40	48·30	0·042
2·13	1·66	98·34	3·32	48·34	0·041
2·14	1·62	98·38	3·24	48·38	0·040
2·15	1·58	98·42	3·16	48·42	0·040
2·16	1·54	98·46	3·08	48·46	0·039
2·17	1·50	98·50	3·00	48·50	0·038
2·18	1·46	98·54	2·93	48·54	0·037
2·19	1·43	98·57	2·85	48·57	0·036
2·20	1·39	98·61	2·78	48·61	0·035
2·21	1·36	98·64	2·71	48·64	0·035
2·22	1·32	98·68	2·64	48·68	0·034
2·23	1·29	98·71	2·58	48·71	0·033
2·24	1·26	98·74	2·51	48·74	0·032
2·25	1·22	98·78	2·45	48·78	0·032
2·26	1·19	98·81	2·38	48·81	0·031
2·27	1·16	98·84	2·32	48·84	0·030
2·28	1·13	98·87	2·26	48·87	0·030
2·29	1·10	98·90	2·20	48·90	0·029
2·30	1·07	98·93	2·15	48·93	0·028
2·31	1·05	98·95	2·09	48·95	0·028
2·32	1·02	98·98	2·04	48·98	0·027
2·33	**0·99**	**99·01**	**1·98**	**49·01**	**0·026**
2·34	0·97	99·03	1·93	49·03	0·026
2·35	0·94	99·06	1·88	49·06	0·025
2·36	0·91	99·09	1·83	49·09	0·025
2·37	0·89	99·11	1·78	49·11	0·024
2·38	0·87	99·13	1·73	49·13	0·023
2·39	0·84	99·16	1·69	49·16	0·023
2·40	0·82	99·18	1·64	49·18	0·022
2·41	0·80	99·20	1·60	49·20	0·022
2·42	0·78	99·22	1·55	49·22	0·021

Table 2.7.2—continued.

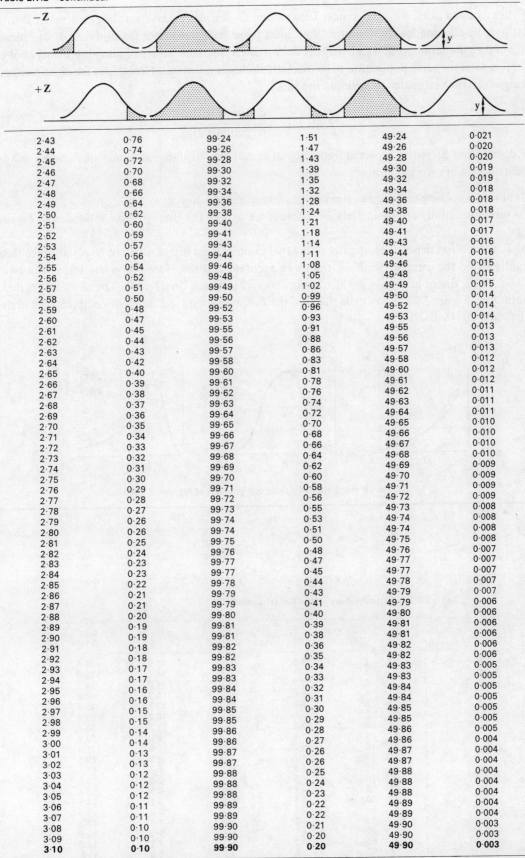

2·43	0·76	99·24	1·51	49·24	0·021
2·44	0·74	99·26	1·47	49·26	0·020
2·45	0·72	99·28	1·43	49·28	0·020
2·46	0·70	99·30	1·39	49·30	0·019
2·47	0·68	99·32	1·35	49·32	0·019
2·48	0·66	99·34	1·32	49·34	0·018
2·49	0·64	99·36	1·28	49·36	0·018
2·50	0·62	99·38	1·24	49·38	0·018
2·51	0·60	99·40	1·21	49·40	0·017
2·52	0·59	99·41	1·18	49·41	0·017
2·53	0·57	99·43	1·14	49·43	0·016
2·54	0·56	99·44	1·11	49·44	0·016
2·55	0·54	99·46	1·08	49·46	0·015
2·56	0·52	99·48	1·05	49·48	0·015
2·57	0·51	99·49	1·02	49·49	0·015
2·58	0·50	99·50	0·99	49·50	0·014
2·59	0·48	99·52	0·96	49·52	0·014
2·60	0·47	99·53	0·93	49·53	0·014
2·61	0·45	99·55	0·91	49·55	0·013
2·62	0·44	99·56	0·88	49·56	0·013
2·63	0·43	99·57	0·86	49·57	0·013
2·64	0·42	99·58	0·83	49·58	0·012
2·65	0·40	99·60	0·81	49·60	0·012
2·66	0·39	99·61	0·78	49·61	0·012
2·67	0·38	99·62	0·76	49·62	0·011
2·68	0·37	99·63	0·74	49·63	0·011
2·69	0·36	99·64	0·72	49·64	0·011
2·70	0·35	99·65	0·70	49·65	0·010
2·71	0·34	99·66	0·68	49·66	0·010
2·72	0·33	99·67	0·66	49·67	0·010
2·73	0·32	99·68	0·64	49·68	0·010
2·74	0·31	99·69	0·62	49·69	0·009
2·75	0·30	99·70	0·60	49·70	0·009
2·76	0·29	99·71	0·58	49·71	0·009
2·77	0·28	99·72	0·56	49·72	0·009
2·78	0·27	99·73	0·55	49·73	0·008
2·79	0·26	99·74	0·53	49·74	0·008
2·80	0·26	99·74	0·51	49·74	0·008
2·81	0·25	99·75	0·50	49·75	0·008
2·82	0·24	99·76	0·48	49·76	0·007
2·83	0·23	99·77	0·47	49·77	0·007
2·84	0·23	99·77	0·45	49·77	0·007
2·85	0·22	99·78	0·44	49·78	0·007
2·86	0·21	99·79	0·43	49·79	0·007
2·87	0·21	99·79	0·41	49·79	0·006
2·88	0·20	99·80	0·40	49·80	0·006
2·89	0·19	99·81	0·39	49·81	0·006
2·90	0·19	99·81	0·38	49·81	0·006
2·91	0·18	99·82	0·36	49·82	0·006
2·92	0·18	99·82	0·35	49·82	0·006
2·93	0·17	99·83	0·34	49·83	0·005
2·94	0·17	99·83	0·33	49·83	0·005
2·95	0·16	99·84	0·32	49·84	0·005
2·96	0·16	99·84	0·31	49·84	0·005
2·97	0·15	99·85	0·30	49·85	0·005
2·98	0·15	99·85	0·29	49·85	0·005
2·99	0·14	99·86	0·28	49·86	0·005
3·00	0·14	99·86	0·27	49·86	0·004
3·01	0·13	99·87	0·26	49·87	0·004
3·02	0·13	99·87	0·26	49·87	0·004
3·03	0·12	99·88	0·25	49·88	0·004
3·04	0·12	99·88	0·24	49·88	0·004
3·05	0·12	99·88	0·23	49·88	0·004
3·06	0·11	99·89	0·22	49·89	0·004
3·07	0·11	99·89	0·22	49·89	0·004
3·08	0·10	99·90	0·21	49·90	0·003
3·09	0·10	99·90	0·20	49·90	0·003
3·10	**0·10**	**99·90**	**0·20**	**49·90**	**0·003**

2.7.3 Fisher's F distribution

The F distributions deals with differences between small samples taken from large normally distributed populations. Small random samples are often used to provide estimates (section 1.5) of the mean and variance of the population from which they are drawn. Because of the nature of sampling we may not expect two samples from the same population to give exactly the same estimates. F measures the difference between two sample variance estimates by forming the ratio:

$$F_{df(1), df(2)} = \frac{\hat{\sigma}^2 \text{ (sample 1)}}{\hat{\sigma}^2 \text{ (sample 2)}} \tag{2.7.3}$$

The F distribution gives the expected frequency of occurrence of different values of F. Some features of the F distribution are worthy of note:

(a) F is always a positive number (a ratio of 2 positive numbers).
(b) F is most commonly approximately 1·0 because we expect the two variance estimates to be roughly equal.
(c) The F distribution depends on the size of the two samples used (these need not be equal) since the more broadly based the sample estimates, the more accurate and therefore the more similar the estimates will be. The F distribution has a wider application when based on degrees of freedom (df) rather than sample size. Figure 2.7.3a shows the shape of the F distribution for 3 different combinations of degrees of freedom (df(1), df(2)).

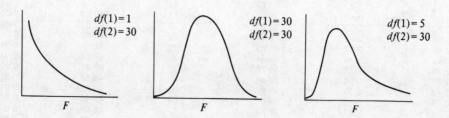

Fig. 2.7.3a **F distributions for three combinations of degrees of freedom.**

Table 2.7.3   **Critical values of the F distribution.**

		2-TAIL: 10% / 1-TAIL: 5%	5% / 2·5%	2% / 1%	0·2% / 0·1%
$df_1 = 1$	$df_2 = 2$	$F \geq 18.5$	38·5	98·5	998·5
1	3	10·1	17·4	34·1	176·0
1	4	7·7	12·2	21·2	74·1
1	5	6·6	10·0	16·3	47·2
1	6	6·0	8·8	13·7	35·5
1	7	6·0	8·1	12·2	29·2
1	8	5·3	7·6	11·3	25·4
1	9	5·1	7·2	10·6	22·9
1	10	5·0	6·9	10·0	21·0
1	12	4·7	6·5	9·3	18·6
1	14	4·6	6·3	8·7	17·1
1	16	4·5	6·1	8·5	16·1
1	18	4·4	6·0	8·3	15·4
1	20	4·3	5·9	8·1	14·8
1	25	4·2	5·7	7·8	13·9
1	30	4·2	5·6	7·6	13·3
1	40	4·1	5·4	7·3	12·6
1	50	4·0	5·3	7·2	12·2
1	100	3·9	5·2	6·9	11·5
1	200	3·9	5·1	6·8	11·2

Table 2.7.3 deals with the situation where the following assumptions are met:

(i) Samples 1 and 2 were both drawn at random.
(ii) The samples were drawn from the same population (or identical populations).
(iii) The parent population was normally distributed.

Large values of F, however, are often taken as indications that at least one of these assumptions has not been met. When assumptions (i) and (iii) can be shown to be true, large values of F are assumed to show that the samples were drawn from populations with different variances.

Table 2.7.3 shows values of F which are exceeded only on a given percentage of occasions when the above assumptions are met. These tables apply when the first sample estimate is used for the numerator of the F ratio. If we wish to use the *larger* of the two estimates:

$$F = \frac{\hat{\sigma}^2 \text{ (sample 1)}}{\hat{\sigma}^2 \text{ (sample 2)}} \quad or \quad \frac{\hat{\sigma}^2 \text{ (sample 2)}}{\hat{\sigma}^2 \text{ (sample 1)}}$$

to obtain an F value greater than 1·0, then there are *twice* as many examples of such values of F. In such circumstances, we must *double* the percentage frequency given for a 1-TAIL test. This procedure is referred to as a 2-TAIL test because we are, in effect, considering very small values of F as well as very large values.

We use a 2-TAIL test whenever we decide *after the results are available* which sample estimate will form the numerator of the F ratio.

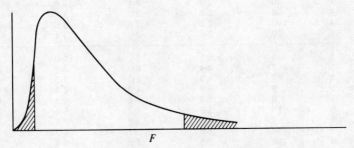

Fig. 2.7.3b   Illustration of a 2-TAIL F test.

Table 2.7.3—continued.

| | | 2-TAIL | 10% | 5% | 2% | 0·2% |
		1-TAIL	5%	2·5%	1%	0·1%
2	3		9·5	16·0	30·8	148·5
2	4		6·9	10·6	18·0	61·2
2	5		5·8	8·4	13·3	37·1
2	6		5·1	7·3	10·9	27·0
2	7		4·7	6·5	9·5	21·7
2	8		4·7	6·1	8·6	18·5
2	9		4·3	5·7	8·0	16·4
2	10		4·1	5·5	7·6	14·9
2	12		3·9	5·1	6·9	13·0
2	14		3·7	4·9	6·5	11·8
2	16		3·6	4·7	6·2	11·0
2	18		3·5	4·6	6·0	10·4
2	20		3·5	4·5	5·8	9·9
2	25		3·4	4·3	5·6	9·2
2	30		3·3	4·2	5·4	8·8
2	40		3·2	4·0	5·2	8·2
2	50		3·2	4·0	5·1	8·0
2	100		3·1	3·8	4·8	7·4
2	200		3·0	3·8	4·7	7·2
$df_1 = 3$	$df_2 = 3$	$F \geq 9·3$		15·4	29·4	140·8
3	4	6·6		10·0	16·7	56·3
3	5	5·4		7·8	12·1	33·2
3	6	4·8		6·6	9·8	23·7
3	7	4·3		5·9	8·5	18·8
3	8	4·1		5·4	7·6	15·8
3	9	3·9		5·1	7·0	13·9

Table 2.7.3   Critical values of the F distribution—continued.

		2-TAIL 1-TAIL	10% 5%	5% 2·5%	2% 1%	0·2% 0·1%
3	10		3·7	4·8	6·6	12·6
3	12		3·5	4·5	6·0	10·8
3	14		3·3	4·2	5·6	9·7
3	16		3·2	4·1	5·3	9·0
3	18		3·2	4·0	5·1	8·5
3	20		3·1	3·9	4·9	8·1
3	25		3·0	3·7	4·7	7·5
3	30		2·9	3·6	4·5	7·1
3	40		2·8	3·5	4·3	6·6
3	50		2·8	3·4	4·2	6·3
3	100		2·7	3·3	4·0	5·9
3	200		2·7	3·2	3·9	5·6
4	4		6·4	9·6	16·0	53·5
4	5		5·2	7·4	11·4	31·1
4	6		4·5	6·2	9·1	21·8
4	7		4·1	5·5	7·8	17·2
4	8		3·8	5·1	7·0	14·4
4	9		3·6	4·7	6·4	12·6
4	10		3·5	4·5	6·0	11·3
4	12		3·3	4·1	5·4	9·6
4	14		3·1	3·9	5·0	8·6
4	16		3·0	3·7	4·8	7·9
4	18		2·9	3·6	4·6	7·5
4	20		2·9	3·5	4·4	7·1
4	25		2·8	3·4	4·2	6·5
4	30		2·7	3·3	4·0	6·1
4	40		2·6	3·1	3·8	5·7
4	50		2·6	3·1	3·7	5·5
4	100		2·5	2·9	3·5	5·0
4	200		2·4	2·9	3·4	4·8
$df_1 = 5$	$df_2 = 5$	$F \geq 5·1$		7·1	11·0	29·7
5	6		4·4	6·0	8·7	20·8
5	7		4·0	5·3	7·5	16·2
5	8		3·7	4·8	6·6	13·5
5	9		3·5	4·5	6·1	11·7
5	10		3·3	4·2	5·6	10·5
5	12		3·1	3·9	5·1	8·9
5	14		3·0	3·7	4·7	7·9
5	16		2·9	3·5	4·4	7·3
5	18		2·8	3·4	4·2	6·8
5	20		2·7	3·3	4·1	6·5
5	25		2·6	3·1	3·9	5·9
5	30		2·5	3·0	3·7	5·5
5	40		2·5	2·9	3·5	5·1
5	50		2·4	2·8	3·4	4·9
5	100		2·3	2·7	3·2	4·5
5	200		2·3	2·6	3·1	4·3
6	6		4·3	5·8	8·5	20·0
6	7		3·9	5·1	7·2	15·5
6	8		3·6	4·7	6·4	12·9
6	9		3·4	4·3	5·8	11·1
6	10		3·2	4·1	5·4	9·9
6	12		3·0	3·7	4·8	8·4
6	14		2·8	3·5	4·5	7·4
6	16		2·7	3·3	4·2	6·8
6	18		2·7	3·2	4·0	6·4
6	20		2·6	3·1	3·9	6·0
6	25		2·5	3·0	3·6	5·5
6	30		2·4	2·9	3·5	5·1
6	40		2·3	2·7	3·3	4·7
6	50		2·3	2·7	3·0	4·1
6	100		2·2	2·5	3·0	4·1
6	200		2·1	2·5	2·9	3·9
$df_1 = 7$	$df_2 = 7$	$F \geq 3·8$		5·0	7·0	15·0
7	8		3·5	4·5	6·2	12·4
7	9		3·3	4·2	5·6	10·7
7	10		3·1	3·9	5·2	9·5
7	12		2·9	3·6	4·6	8·0
7	14		2·8	3·4	4·3	7·1
7	16		2·7	3·2	4·0	6·5
7	18		2·6	3·1	3·8	6·0
7	20		2·5	3·0	3·7	5·7
7	25		2·4	2·8	3·5	5·1
7	30		2·3	2·7	3·3	4·8

Table 2.7.3—continued.

		2-TAIL 1-TAIL	10% 5%	5% 2·5%	2% 1%	0·2% 0·1%
7	40		2·2	2·6	3·1	4·4
7	50		2·2	2·6	3·0	4·2
7	100		2·1	2·4	2·8	3·8
7	200		2·1	2·4	2·7	3·6
8	8		3·4	4·4	6·0	12·0
8	9		3·2	4·1	5·5	10·4
8	10		3·1	3·9	5·1	9·2
8	12		2·8	3·5	4·5	7·7
8	14		2·7	3·3	4·1	6·8
8	16		2·6	3·1	3·9	6·2
8	18		2·5	3·0	3·7	5·8
8	20		2·4	2·9	3·6	5·4
8	25		2·3	2·8	3·3	4·9
8	30		2·3	2·7	3·2	4·6
8	40		2·2	2·5	3·0	4·2
8	50		2·1	2·5	2·9	4·0
8	100		2·0	2·3	2·7	3·6
8	200		2·0	2·3	2·6	3·4
$df_1 = 9$	$df_2 = 9$	$F \geq 3·2$		4·0	5·4	10·1
9	10		3·0	3·8	4·9	9·0
9	12		2·8	3·4	4·4	7·5
9	14		2·6	3·2	4·0	6·6
9	16		2·5	3·0	3·8	6·0
9	18		2·5	2·9	3·6	5·6
9	20		2·4	2·8	3·5	5·2
9	25		2·3	2·7	3·2	4·7
9	30		2·2	2·6	3·1	4·4
9	40		2·1	2·5	2·9	4·0
9	50		2·1	2·4	2·8	3·8
9	100		2·0	2·2	2·6	3·4
9	200		1·9	2·2	2·5	3·3
10	10		3·0	3·7	4·8	8·8
10	12		2·8	3·4	4·3	7·3
10	14		2·6	3·1	3·9	6·4
10	16		2·5	3·0	3·7	5·8
10	18		2·4	2·9	3·5	5·4
10	20		2·3	2·8	3·4	5·1
10	25		2·2	2·6	3·1	4·6
10	30		2·2	2·5	3·0	4·2
10	40		2·1	2·4	2·8	3·9
10	50		2·0	2·3	2·7	3·7
10	100		1·9	2·2	2·5	3·3
10	200		1·9	2·1	2·4	3·1
$df_1 = 12$	$df_2 = 12$	$F \geq 2·7$		3·3	4·2	7·0
12	14		2·5	3·1	3·8	6·1
12	16		2·4	2·9	3·6	5·5
12	18		2·3	2·8	3·4	5·1
12	20		2·3	2·7	3·2	4·8
12	25		2·2	2·5	3·0	4·3
12	30		2·1	2·4	2·8	4·0
12	40		2·0	2·3	2·7	3·6
12	50		2·0	2·2	2·6	3·4
12	100		1·8	2·1	2·4	3·1
12	200		1·8	2·0	2·3	2·9
20	20		2·1	2·5	2·9	4·3
20	25		2·0	2·3	2·7	3·8
20	30		1·9	2·2	2·5	3·5
20	40		1·8	2·1	2·4	3·1
20	50		1·8	2·0	2·3	3·0
20	100		1·7	1·8	2·1	2·6
20	200		1·6	1·8	2·0	2·4
50	50		1·6	1·8	1·9	2·4
50	100		1·5	1·6	1·7	2·1
50	200		1·4	1·5	1·6	1·9
100	100		1·4	1·5	1·6	1·9
100	200		1·3	1·4	1·5	1·7

2.7.4 Chi square distribution

The chi square distribution can be seen as a special case of the F distribution where only one random sample is taken and where the true variance of the population is known (σ^2). In this case, the F ratio is:

$$F_{df(1),\infty} = \frac{\hat{\sigma}^2 \text{ (sample)}}{\sigma^2 \text{ (population)}} = \chi^2/df(1)$$

from which we have

$$\chi^2_{df(1)} = \frac{df(1) \times \hat{\sigma}^2 \text{ (sample)}}{\sigma^2 \text{ (population)}} \qquad (2.7.4a)$$

where:

$\hat{\sigma}^2$ (sample) is an estimate of the population variance based on a random sample.

$df(1)$ is the number of degrees of freedom associated with the sample.

σ^2 is the true variance of the population from which the sample was drawn. (For the purposes of the comparison with the F ratio, σ^2 is assumed to have an infinite (∞) number of degrees of freedom.)

The chi square distribution gives the expected frequency of occurrence of different values of χ^2 when our sample is drawn in a random fashion from a very large normally distributed population. Some features of the χ^2 distribution are worthy of note:

(a) χ^2 is always a positive number.

(b) χ^2 is most commonly approximately equal to $df(1)$ because we expect the ratio of the variance estimate to the true variance to be approximately 1·0.

(c) The chi square distribution depends upon the size of the sample since, the more broadly based the sample, the more accurate the variance estimate will be. Figure 2.7.4 shows the shape of the chi square distribution for three different degrees of freedom.

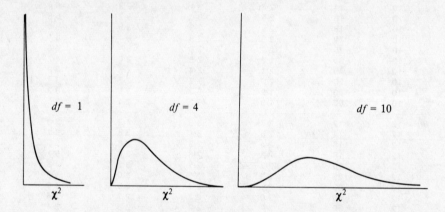

Fig. 2.7.4   Chi square distribution for three different values of df.

Table 2.7.4   Critical values of the χ^2 distribution.

Significance level (1-TAIL)	5%	2·5%	1%	0·1%
$df = 1$	$\chi^2 \geq 3{\cdot}9$	5·0	6·6	10·8
2	6·0	7·4	9·2	13·8
3	7·8	9·3	11·3	16·3
4	9·5	11·1	13·3	18·5
5	11·1	12·8	15·1	20·5
6	12·6	14·5	16·8	22·5
7	14·1	16·0	18·5	24·3
8	15·5	17·5	20·1	26·1
9	16·9	19·0	21·7	27·9
10	18·3	20·5	23·2	29·6
11	19·7	21·9	24·7	31·3
12	21·0	23·3	26·2	32·9
13	22·4	24·7	27·7	34·5
14	23·7	26·1	29·1	36·1
15	25·0	27·5	30·6	37·7
16	26·3	28·8	32·0	39·3
17	27·6	30·2	33·4	40·8
18	28·9	31·5	34·8	42·3
19	30·1	32·9	36·2	43·8
20	31·4	34·2	37·6	45·3
25	37·7	40·6	44·3	42·6
30	43·8	47·0	50·9	59·7
40	55·8	59·3	63·7	73·4
50	67·5	71·4	76·2	86·7
60	79·1	83·3	88·4	99·6

Table 2.7.4 is based on the situation where the following assumptions are met:

 (i) The sample was drawn at random.

 (ii) The sample was drawn from a population whose true variance (σ^2) is specified in the calculation of χ^2.

(iii) The parent population was normally distributed.

   Large values of χ^2, however, are often taken as indications that at least one of these assumptions has not been met. When assumptions (i) and (ii) can be shown to be true, large values of χ^2 imply that assumption (i) is not true. This implies that the variance of the population from which the sample was taken is not, in fact, what it was thought to be, i.e., the value σ^2 which was used in the calculation of χ^2 (formula 2.7.4) was probably *not* the true value of the population variance.

   Table 2.7.4 gives large values of χ^2 which are exceeded on only a given percentage of occasions when the above assumptions are met. Only extreme large values (not extreme small values) are considered because most of the applications in use by non-statisticians refer only to larger values of χ^2. Accordingly 2-TAIL test considerations do not normally apply.

   For large numbers of degrees of freedom, convert χ^2 to z which can be assessed by reference to table 2.7.2

$$z = \sqrt{2\chi^2} - \sqrt{(2df - 1)} \tag{2.7.4b}$$

For convenience, some critical values of z are given below

1-TAIL	5%	2·5%	1%	0·1%
$Z \geq$	1·64	1·96	2·33	3·10

2.7.5 Student's t distribution

Like the chi square distribution, Student's t distribution can also be viewed as a special case of the F distribution. We again compare two variance estimates based upon random samples, but on this occasion the first sample has only 1 degree of freedom.

$$F_{1, df(2)} = \frac{\hat{\sigma}_1^2}{\hat{\sigma}_2^2} = \frac{d^2}{\hat{\sigma}_2^2} = t_{df(2)}^2 \qquad (2.7.5a)$$

where:
$d^2 = (X - \mu)^2$,
$\hat{\sigma}_2^2$ is based on a sample with $df(2)$ degrees of freedom, and
X is a single score drawn at random from a population with known mean, μ.

It follows that:

$$t_{df(2)} = \frac{d}{\hat{\sigma}_2} = \sqrt{F_{1, df(2)}} \qquad (2.7.5b)$$

The t distribution gives the expected frequency of occurrence of different values of t. Some features of the t distribution should be noted:

(a) t may be positive or negative (if X is less than μ, d is negative).
(b) t is most commonly approximately zero because we do not expect X to be much different from μ.
(c) The t distribution depends on the size of the large sample since the bigger the sample the more accurate the value of $\hat{\sigma}_2^2$. When df is very large, t is approximately normally distributed.

Table 2.7.5 deals with the situation where the following assumptions are met:

(i) Both the single score (X) and the sample of scores were drawn at random.
(ii) Both the single score (X) and the sample were drawn from the same population.
(iii) The parent population was normally distributed.

Table 2.7.5 Critical values of t for Student's t distribution.

2-TAIL 1-TAIL	10% 5%	5% 2·5%	2% 1%	1% 0·1%
$df = 2$	$t \geq 2·92$	4·30	6·96	22·33
3	2·35	3·18	4·54	10·21
4	2·13	2·78	3·75	7·17
5	2·02	2·57	3·36	5·89
6	1·94	2·45	3·14	5·21
7	1·89	2·36	3·00	4·79
8	1·86	2·31	2·90	4·50
9	1·83	2·26	2·82	4·30
10	1·81	2·23	2·76	4·14
11	1·80	2·20	2·72	4·02
12	1·78	2·18	2·68	3·93
13	1·77	2·16	2·65	3·85
14	1·76	2·14	2·62	3·79
15	1·75	2·13	2·60	3·73
16	1·75	2·12	2·58	3·69
17	1·73	2·10	2·57	3·65
18	1·73	2·10	2·55	3·61
19	1·73	2·09	2·54	3·58
20	1·72	2·09	2·53	3·55
25	1·71	2·06	2·49	3·45
30	1·70	2·04	2·46	3·39
40	1·68	2·02	2·42	3·31
50	1·68	2·01	2·40	3·26
60	1·67	2·00	2·39	3·23

Extreme values of t are often taken as indications that one of these three assumptions is not true. When assumptions (i) and (iii) can be shown to be true, extreme values of t imply that the score (X) was not drawn from the same population as the sample of scores.

Table 2.7.2 gives values of t which are exceeded on a given percentage of occasions when the assumptions are met. The 1-TAIL frequency values concern *only positive* values of t. The 2-TAIL frequency values concern *both* positive and negative values of t. 1-TAIL frequency values are appropriate when the sign (positive or negative) of t was correctly indicated before the results were available. The t distribution is symmetrical. Accordingly, table 2.7.2 can be used to assess the significance of negative values of t.

2.7.6 Poisson distribution

The Poisson distribution deals with very rare events which are detected in samples which are very large. The sample size is so large as to be irrelevant. The important measure for each sample is K, the number of events detected in a single sample. The Poisson distribution is entirely determined by μ, the average number of these events expected in a sample. Figure 2.7.6 illustrates the Poisson frequency distributions for three different values of μ.

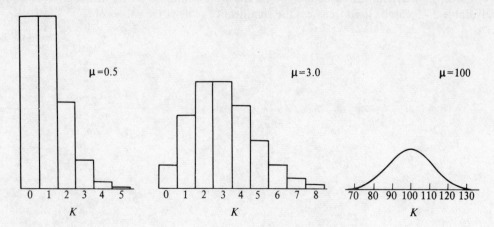

Fig. 2.7.6 Poisson distribution for three values of μ.

Table 2.7.6 gives some exact frequency distributions for various values of μ. Values in the body of the table give the percentage frequencies of various values of K. When the mean of a Poisson distribution is known (μ) we may immediately find the variance:

$$\sigma^2 = \mu \tag{2.7.6a}$$

The Poisson distribution comes to *approximate the normal distribution as μ increases* (see Fig. 2.7.6).

When μ is small, it is often necessary to compute the exact percentage frequency of a given value of K.

$$P_{(\mu,k)} = 100\,\frac{\mu^k}{K!\,e^\mu} \tag{2.7.6b}$$

where:

μ is the mean of a Poisson distribution.
K is the number of rare events detected in the sample
$P_{(\mu,k)}$ is the percentage frequency of samples containing K rare events.
e is $2\cdot718\ldots$
$K!$ is 'K factorial' $= K(K-1)(K-2)\ldots3.2.1$

A convenient computational formula for calculating $P_{(\mu,k)}$ is:

$$\log P_{(\mu,k)} = 2 + k\log_{10}(\mu) - \log_{10}(K!) - 0\cdot4343\mu \tag{2.7.6c}$$

Table 2.7.6 Poisson percentage frequency distributions.

μ =	0	1	2	3	4	5	6	7	8	9	10	11	12	13	14	15	16	17	18	19	20
0.2	81.9	16.4	1.6	0.1																	
0.4	67.0	26.8	5.4	0.7	0.1																
0.6	54.9	32.9	9.9	2.0	0.3																
0.8	44.9	35.9	14.4	3.8	0.8	0.1															
1.0	36.8	36.8	18.4	6.1	1.5	0.3	0.1														
1.2	30.1	36.1	21.7	8.7	2.6	0.6	0.1														
1.4	24.7	34.5	24.2	11.3	3.9	1.1	0.3	0.1													
1.6	20.2	32.3	25.8	13.8	5.5	1.8	0.5	0.1													
1.8	16.5	29.8	26.8	16.1	7.2	2.6	0.8	0.2													
2.0	13.5	27.1	27.1	18.0	9.0	3.6	1.2	0.3	0.1												
2.2	11.1	24.4	26.8	19.7	10.8	4.8	1.7	0.5	0.2												
2.4	9.1	21.8	26.1	20.9	12.5	6.0	2.4	0.8	0.2	0.1											
2.6	7.4	19.3	25.1	21.8	14.1	7.4	3.2	1.2	0.4	0.1											
2.8	6.1	17.0	23.8	22.2	15.6	8.7	4.1	1.6	0.6	0.2											
3.0	5.0	14.9	22.4	22.4	16.8	10.1	5.0	2.2	0.8	0.3	0.1										
3.2	4.1	13.0	20.9	22.3	17.8	11.4	6.1	2.8	1.1	0.4	0.1										
3.4	3.3	11.3	19.3	21.9	18.6	12.6	7.2	3.5	1.5	0.6	0.2	0.1									
3.6	2.7	9.8	17.7	21.2	19.1	13.8	8.3	4.2	1.9	0.8	0.3	0.1									
3.8	2.2	8.5	16.2	20.5	19.4	14.8	9.4	5.1	2.4	1.0	0.4	0.1									
4.0	1.8	7.3	14.7	19.5	19.5	15.6	10.4	6.0	3.0	1.3	0.5	0.2	0.1								
4.2	1.5	6.3	13.2	18.5	19.4	16.3	11.4	6.9	3.6	1.7	0.7	0.3	0.1								
4.4	1.2	5.4	11.9	17.4	19.2	16.9	12.4	7.8	4.3	2.1	0.9	0.4	0.1								
4.6	1.0	4.6	10.6	16.3	18.8	17.3	13.2	8.7	5.0	2.6	1.2	0.5	0.2	0.1							
4.8	0.8	4.0	9.5	15.2	18.2	17.5	14.0	9.6	5.8	3.1	1.5	0.6	0.3	0.1							
5.0	0.7	3.4	8.4	14.0	17.5	17.5	14.6	10.4	6.5	3.6	1.8	0.8	0.3	0.1							
5.2	0.6	2.9	7.5	12.9	16.8	17.5	15.1	11.3	7.3	4.2	2.2	1.0	0.5	0.2	0.1						
5.4	0.5	2.4	6.6	11.9	16.0	17.3	15.6	12.0	8.1	4.9	2.6	1.3	0.6	0.2	0.1						
5.6	0.4	2.1	5.8	10.8	15.2	17.0	15.8	12.7	8.9	5.5	3.1	1.6	0.7	0.3	0.1						
5.8	0.3	1.8	5.1	9.8	14.3	16.6	16.0	13.3	9.6	6.2	3.6	1.9	0.9	0.4	0.2	0.1					
6.0	0.2	1.5	4.5	8.9	13.4	16.1	16.1	13.8	10.3	6.9	4.1	2.3	1.1	0.5	0.2	0.1					
6.2	0.2	1.3	3.9	8.1	12.5	15.5	16.0	14.2	11.0	7.6	4.7	2.6	1.4	0.7	0.3	0.1					
6.4	0.2	1.1	3.4	7.3	11.6	14.9	15.9	14.5	11.6	8.2	5.3	3.1	1.6	0.8	0.4	0.2	0.1				
6.6	0.1	0.9	3.0	6.5	10.7	14.2	15.6	14.7	12.1	8.9	5.9	3.5	1.9	1.0	0.5	0.2	0.1				
6.8	0.1	0.8	2.6	5.8	9.9	13.5	15.3	14.9	12.6	9.5	6.5	4.0	2.3	1.2	0.6	0.3	0.1				
7.0	0.1	0.6	2.2	5.2	9.1	12.8	14.9	14.9	13.0	10.1	7.1	4.5	2.6	1.4	0.7	0.3	0.1	0.1			
7.2	0.1	0.5	1.9	4.6	8.4	12.0	14.4	14.9	13.4	10.7	7.7	5.0	3.0	1.7	0.9	0.4	0.2	0.1			
7.4	0.1	0.5	1.7	4.1	7.6	11.3	13.9	14.7	13.6	11.2	8.3	5.6	3.4	2.0	1.0	0.5	0.2	0.1			
7.6	0.1	0.4	1.4	3.7	7.0	10.6	13.4	14.5	13.8	11.7	8.9	6.1	3.9	2.3	1.2	0.6	0.3	0.1	0.1		
7.8	0.0	0.3	1.2	3.2	6.3	9.9	12.8	14.3	13.9	12.1	9.4	6.7	4.3	2.6	1.5	0.8	0.4	0.2	0.1		
8.0	0.0	0.3	1.1	2.9	5.7	9.2	12.2	14.0	14.0	12.4	9.9	7.2	4.8	3.0	1.7	0.9	0.5	0.2	0.1		
8.2	0.0	0.2	0.9	2.5	5.2	8.5	11.6	13.6	13.9	12.7	10.4	7.8	5.3	3.3	2.0	1.1	0.5	0.3	0.1	0.1	
8.4	0.0	0.2	0.8	2.2	4.7	7.8	11.0	13.2	13.8	12.9	10.8	8.3	5.8	3.7	2.2	1.3	0.7	0.3	0.2	0.1	
8.6	0.0	0.2	0.7	2.0	4.2	7.2	10.3	12.7	13.7	13.1	11.2	8.8	6.3	4.2	2.6	1.5	0.8	0.4	0.2	0.1	
8.8	0.0	0.1	0.6	1.7	3.8	6.6	9.7	12.2	13.4	13.1	11.6	9.3	6.8	4.6	2.9	1.7	0.9	0.5	0.2	0.1	
9.0	0.0	0.1	0.5	1.5	3.4	6.1	9.1	11.7	13.2	13.2	11.9	9.7	7.3	5.0	3.2	1.9	1.1	0.6	0.3	0.1	0.1
9.2	0.0	0.1	0.4	1.3	3.0	5.5	8.5	11.2	12.9	13.1	12.1	10.1	7.8	5.5	3.6	2.2	1.3	0.7	0.4	0.2	0.1
9.4	0.0	0.1	0.4	1.1	2.7	5.1	7.9	10.6	12.5	13.1	12.3	10.5	8.2	5.9	4.0	2.5	1.5	0.8	0.4	0.2	0.1
9.6	0.0	0.1	0.3	1.0	2.4	4.6	7.4	10.1	12.1	12.9	12.4	10.8	8.7	6.4	4.4	2.8	1.7	1.0	0.5	0.3	0.1
9.8	0.0	0.1	0.3	0.9	2.1	4.2	6.8	9.5	11.7	12.7	12.5	11.1	9.1	6.8	4.8	3.1	1.9	1.1	0.6	0.3	0.2
10.0	0.0	0.0	0.2	0.8	1.9	3.8	6.3	9.0	11.3	12.5	12.5	11.4	9.5	7.3	5.2	3.5	2.2	1.3	0.7	0.4	0.2
	K = 0	1	2	3	4	5	6	7	8	9	10	11	12	13	14	15	16	17	18	19	20

3. Choosing a test

This chapter provides a scheme which, if followed carefully, should help pinpoint the required test quickly. The scheme assumes that the data can be analysed using the tests in this book. If the test is not included in the book then hopefully the scheme will show this too and you will not be misled into using an inappropriate test.

To use the scheme, answer the following questions:

1. What questions do you intend to ask about the data?
2. How many *variables* does your question concern? Section 3.1 helps you identify the variables. Most simple statistical tests involve two variables.
3. What type of scores is used by your variables? Section 3.2 helps you identify the type of scores.

If you have only *1 variable*, table 3.3 will help choose your test according to the scale used. If you have *only 2 variables*, you must decide whether your scores are *related* or *unrelated*. Section 3.2.4 helps you decide appropriately. Tests for *unrelated* scores are summarized in table 3.4.1 according to the scales of your 2 variables. Tests for *related* scores are summarized in table 3.4.2. If you have *more than 2 variables* you may be straying into the deeper waters of analysis of variance, see section 3.5. If you have exactly 3 variables, however, you should carefully consider the possibility that your question really concerns only 2 variables with related scores. This is especially likely when one of your variables is *subjects*.

The scheme will be illustrated using the following experimental set-up. This example concerns a course in introductory statistics given to approximately 300 students from many different disciplines (e.g., Botany, Zoology, Physiology, Psychology, Education, Sociology, Economics, Business Studies). For many years, the organizers of the course have been aware that the course was unsatisfactory in many ways. Students from some departments responded well to the course while others found it irrelevant and difficult. The organizers considered the possibility that only students with good abstract reasoning ability were responding to the course material and presentation. To check this idea, all students about to begin the course were given a test of abstract reasoning ability supplied by the Psychology department. All students are given an end of course examination which is used as a measure of success in negotiating the course. Although the examiners mark the papers with a percentage score, these results are only made public as a decile score. (On the decile scale the *worst* 10 per cent of students are given the score 1 while the *best* 10 per cent are given a score of 10. Intermediate students are given scores between 1 and 10 according to how well they fared *relative to their classmates*.)

3.1 How many variables?

In simple terms, variables are properties of objects, events, people, etc. The introductory statistics course could be good or bad. The 'goodness' of the course is then a variable. Students can be described in terms of the department they came from (Botany, Zoology, etc.), therefore 'department' is a variable. Many variables are obvious, e.g., examination results and abstract reasoning test results.

Consider your questions about the data carefully and identify the variables clearly. When we know how many variables are involved we will be able to narrow the range of choice of statistical test.

▶ Example.   Consider the following questions and identify the variables involved.

(a) 'Are the results of the reasoning test normally distributed?'
   This question involves only *one* variable, reasoning ability.

(b) 'Is the course enjoyed most by intelligent students?'
   This question involves *two* variables:

   (i) Enjoyment of the course.
   (ii) Intelligence of the student.

(c) 'Do students in the Biology department work harder at statistics than Sociology students?'
   Two variables:

   (i) Department (Biology/Sociology).
   (ii) Degree of application to statistics.

(d) 'Brighter students are more successful in the end of course examination but this effect is most especially marked in female students.'
   Three variables:

   (i) Intelligence.
   (ii) Sex.
   (iii) Examination results.

3.2 What types of scores?

When a variable is given a numerical value we shall call it a score. There are many types of scores but some are of particular interest.

Nominal scores.   As the name suggests, these scores are merely names.

▶ Examples.   Department: Zoology, Botany, Education, Sociology, etc. Sex: male, female.

Sometimes other types of scores are combined to form only two categories of scores; we may now treat these as nominal.

▶ Example.   Examination result: pass, fail.

Rankable scores.   When the scores can be ordered with respect to one another in a unique fashion we say that the scores are rankable because we could, if we wanted to, assign ranks to the scores and use the ranks instead of the scores in our analysis. Any set of three scores (a, b and c) must have this simple property:

$$\text{If } a < b \text{ and if } b < c \text{ then } a < c$$

▶ Examples.   Examination result: fail, satisfactory, good, or deciles (1, 2, 3, 4, 5, 6, 7, 8, 9, 10). Reasoning ability: below average, average, above average, outstanding.

Rankable categories.   It is possible for a number of people to have the same score; these are called tied scores, see section 2.3. When many people share scores it is often convenient to call the scores *rankable categories.*

Normally distributed scores.   Section 2.7.2 describes the normal distribution. It occurs very often that samples collected in research are drawn from populations of normally distributed scores. When we know that the population is normally distributed, this constitutes useful information which can help make our statistical testing more sensitive. It should be clearly noted that we are concerned with the normality of the *population* rather than the sample. It is not always easy to decide whether the population is normally distributed but the sample distribution can be used as a guide. The decision is rarely very critical, however, and small departures from normality do not matter very much.

Scores which are normally distributed will also be rankable. This means that we can ignore the property of normal distribution and use less sensitive rank order tests where convenient. The reverse is not true: rankable scores are not necessarily normally distributed.

▶ Examples.   We might reasonably expect that the abstract reasoning score (points) and the end of course examination result (percentage) will be approximately normally distributed. This expectation is based on past experience.

Related/unrelated scores. When a random sample is made, all the scores in the sample are by definition independent of each other. When two samples are made, the scores in the two samples may either be *unrelated* or *related*. If the second sample is made completely at random then the scores in one sample must be independent of those in the other. Often, however, the second sample is influenced by the first. For example, a single object or subject may feature in both samples (a) before and (b) after a certain treatment. When this happens, the scores are said to be related. Related scores can usually be identified easily because each score in one sample (or group) has a corresponding score in the second sample. These scores are called *matched pairs*. When there are more than two samples, the related scores are called *matched sets*.

Related scores involve different statistical tests from independent scores. Once you have decided whether the scores are related or independent you have narrowed your choice of statistical test.

▶ Example.
Question: Do women obtain better examination results than men?

1. Experimental design using *unrelated* scores. This involves a *random sample* of men and a *random sample* of women.

	Men	Women	
	m	*m*	
	m	*m*	where *m* indicates a mark
	m	*m*	
	m	*m*	
	etc.		

2. Experimental design using *related* scores. Here each man is *paired* with a woman having an *equal age and intelligence*. Within this restriction, they are selected at random.

	Men	Women
Pair 1	*m*	*m*
Pair 2	*m*	*m*
Pair 3	*m*	*m*
	etc.	

3.3 Tests for one variable

These tests ask whether a collection of scores is distributed in a certain way. To be more precise, we want to know about the population of scores from which our sample was drawn.

When the scores are rankable we may ask whether our scores were drawn from a normal, Poisson or uniform (flat) distribution. When the scores are merely nominal we may still ask whether our sample was drawn from, for example, a population with an equal number of scores in each category or some other specified distribution of the scores across categories.

Table 3.3 names the appropriate test according to the scale in use. Chapter 4 describes and illustrates their use.

Table 3.3   Tests for scores on *one* variable.

| VARIABLE | NOMINAL | | RANKABLE | |
	2-Categories	2 or more categories	Categories	Scores
TEST	Binomial (4.2)	Chi square (4.3)	K–S (4.4)	K–S (4.5)

3.4 Tests for two variables

Tests for two variables are usually tests of association. They come in many disguises but essentially they are intended to ask whether a knowledge of a person's (or object's) score on variable *A* will help us predict his score on variable *B*. We will illustrate this point by asking whether abstract reasoning test performance can help predict end of course examination results. We will ask the same question using many different tests to illustrate how the test is dictated by the nature of the scores used.

For example, the examination result can be reported using three different types of score: (a) Nominal (pass/fail), (b) Rankable (decile scale) or (c) Normally distributed (percentage mark). (The percentage mark will be tolerably rather than exactly normally distributed.) A different test will be appropriate for each type of score.

Table 3.4.1 must be used when the scores are *unrelated*.
Table 3.4.2 must be used when the scores are *related*, i.e., when they involve matched pairs or matched sets of scores. This most commonly occurs when the same subject has been used more than once in an experiment.

3.5 Tests for three or more variables

Such tests commonly involve questions concerning the association between a single (dependent) variable and any number of other (independent) variables. All such tests are basically similar but two groups of tests may be distinguished.

(a) *Analysis of variance* (ANOVAR).   These tests apply where the scores on the *independent* variables form nominal or rankable groups while the *dependent* variable has normally distributed scores. ANOVAR techniques have proved very flexible and many computational procedures have been devised for different experimental designs (trend analysis, planned comparisons, Latin squares, etc.). These procedures generalize to any number of independent variables. See chapter 7.

(b) *Multiple correlation and regression.*   As tests of significance they assume that scores on all variables are normally distributed. The procedures are computationally very tedious when the number of variables is more than three but increasingly they are being handled by computers. These procedures are not described in this book.

Table 3.4.1 Tests for unrelated scores on two variables.

| | | VARIABLE 1 | | | | |
| | | Nominal | | Rankable | | Normally distributed scores |
VARIABLE 2		2 Groups	3 or more groups	Groups	Scores	
Nominal	2 Groups	Fisher's exact (5.2)	Chi square (5.3)	K–S (5.4)	Mann-Whitney (5.5)	Unrelated t (5.6) or ANOVAR (1 factor) (7.3.1)
	3 or more groups		Chi square (5.7)		Kruskal-Wallis (5.8)	ANOVAR (1 factor) (7.3.1)
Rankable	Groups			Kendall's τ (5.9)	Kendall's τ (see 5.9)	ANOVAR (Trend) (7.5.1)
	Scores				Kendall's τ (5.10)	
Normally distributed scores						Pearson's r (8.1)

(ANOVAR = Analysis of variance).

Table 3.4.2 Test for related scores on two variables.

| | | VARIABLE 2 | | | | |
| | | Nominal | | Rankable | | Normally distributed scores |
VARIABLE 1		2 groups	3 or more groups	Groups	Scores	
Nominal	2 Groups	McNemar (6.2)	Cochran's Q (6.3)	Sign test (6.4)	Wilcoxon (6.5) or 2 groups Repeated measures (6.6)	Related t (6.7) or ANOVAR (2 factor) (Chapter 7)
	3 or more groups				Friedman (6.8)	ANOVAR (2 factor) (Chapter 7)
Rankable	Groups				Page's L (6.9)	ANOVAR (trend) (2 factor) (Chapter 7)

(ANOVAR = Analysis of variance).

3.6 Significance testing

3.6.1 Making inferences

Each statistical test is based on a set of special assumptions. When we use the test we are asking whether the assumptions of the test are also true of our data. The test does not provide a firm answer 'yes' or 'no', but does indicate how certain we can be in drawing our conclusions.

We can illustrate this using the binomial test. This test is concerned with the situation where we draw a small sample of scores from a very large population which contains only two types of scores (A and B). The test makes the following assumptions:

(i) The sample is drawn at random.

(ii) The population contains an *equal* number of A and B scores.

Clearly, when we draw a random sample from such a population we *expect* to obtain an equal number of A and B scores, but it is also possible to obtain an unequal number. Table 4.2 shows that 5 per cent of such random samples contain 5 or fewer A scores in a sample of 20. It is even possible, though rarer, to obtain no A scores at all in a sample.

If we perform an experiment where individual results have only two possible outcomes (A and B), we may wish to know whether the outcomes are equally likely. If our sample were to contain very few examples of A we might wish to infer that A outcomes are unlikely, i.e., *that A and B outcomes are not equally likely.* The question becomes 'How small a percentage of A scores do we need to find in our sample before we can conclude that A and B are *not* equally likely outcomes?'.

We can, of course, *never* arrive at a certain conclusion. We saw above that some samples have *no A* scores even when A and B scores are equally likely. There is no simple solution to this problem which is central to most statistical decisions. We can, however, look at those sample values which are unlikely to occur when assumptions (i) and (ii) are true and treat these values as an indication that the assumptions are not true. *Significance tables list sample values which occur infrequently when the assumptions are true.* They specify in this book sample values which are exceeded on fewer than (a) 5%, (b) 2·5%, (c) 1% and (d) 0·1% of occasions.

When we have a sample value which appears to be rare (if the assumptions are true), we consult the tables. If the tables confirm our impression that such a sample value would indeed be rare, we may consider the possibility that *at least one of the assumptions of the test is false.* Some care must now be taken in deciding which assumption is false. If we have designed our experiment properly, we will have excluded all possibilities but one. In this example, if we had made our experimental observations at random then assumption (i) is true and it follows that assumption (ii) must be false.

The assumptions for each test are given in the inference section which follows a description of the procedure. These are usually accompanied by some guidance to help decide whether the experimental design permits us to decide which assumption is untrue. The situation varies somewhat from one test to another but the basic situation remains the same.

3.6.2 Significance levels

There can be no firm rules about which significance level permits us to draw the conclusion that one of our assumptions is untrue. The smaller the significance level, the less likely we are to make a mistake. The tables in this book all give the same significance levels. This is merely to reduce confusion; there is nothing magical about them.

A simple meaning can be attached to a significance level statement. A result which is significant at the 5 per cent level is a result which only occurs in 5 per cent of samples *when the assumptions are true.* If we now conclude that the assumptions are false, there is a 5 per cent chance that we are making a mistake. Clearly, the better the significance level the smaller the chance that we will make a mistake.

Throughout this handbook we use the expression 'better than the 5 per cent (or 1 per cent, etc.) significance level'. We do this because the tables have been drawn up so that the values always occur on *less* than 5 per cent (or 1 per cent, etc.) of occasions. This means that tables are *conservative*, that your results are often slightly *more* significant than the tables specify. Tables in other books are often slightly *less* conservative. This will explain some of the discrepancies you will find.

3.6.3 1- and 2-TAIL tests

Many, but not all, of the tests in this handbook use sample values which could be *larger* or *smaller* than the expected value. In the example given above, we could have observed more or fewer occurrences of *A* values than expected. We had a sample of 20 scores and 'expected' 10 *A* values. We might therefore feel that either 20 *A* scores *or* 0 *A* scores constitute equally potent evidence for believing one of the assumptions to be untrue. In this case, we are using a 2-TAIL test since we are prepared to accept evidence from two ends of the scale. A 2-TAIL test takes care of the fact that we are twice as likely to find contrary evidence; it requires that we have more impressive evidence.

Sometimes, we are interested in evidence from only one end of the scale. For example, we may only admit as evidence values of *A* which are less than the expected value, i.e., less than 10. This is a 1-TAIL test. This situation arises when we have predicted that our sample value will lie below (or above) the expected average value. Our prediction must be based on some good argument or previous experience, of course. A 1-TAIL test allows us to use ancillary information to help decide whether the assumptions of the test apply to our data. Such a test should not be used except where such information exists. A 1-TAIL test carries with it a severe disadvantage; if the results go *against your prediction*, you *cannot* then switch to a 2-TAIL test. Your data cannot be used to test the alternative possibility.

The only difference between the two tests is the final significance level. Each table clearly indicates the significance level which applies to each kind of test.

3.7 Choosing between equivalent tests

Often your data is such that it could be tested using a number of different tests. This arises because normally distributed scores are also rankable and rankable scores are also nominal. It follows that normally distributed scores can be treated *either* as normally distributed *or* as rankable or as nominal. In addition, rankable scores can be treated *either* as rankable *or* as nominal (for example, we may give all ranks above 5 a score of *X* or 1 and all ranks below 5 a score of *Y* or 0). Equivalent tests are easily found *in the same row* of tables 3.4.1 and 3.4.2. In such circumstances, there are two general principles to help us decide which test to use: considerations of *effort* and considerations of the *power of the test*.

The *effort* required for the performance of any test is commonly least for nominal tests, more for rank tests and most for tests assuming normally distributed scores. For example, in table 3.4.2 the three tests applicable to three group tests can be ordered in terms of effort:

Chi square,   Kruskal–Wallis,   ANOVAR

In many cases, this simple rule becomes a matter of opinion. Experience will tell which tests you find least effortful.

The *power* of a test is a reflection of its *sensitivity* or its ability to detect significant effects in the data when they are in fact there. The statistical tables associated with each test are based on a number of assumptions. When a statistically significant result occurs, we are advised to infer that at least one of these assumptions has been violated. When we do so, we run the risk of making a special kind of mistake called a type I error.

A type I error occurs when we reject the assumptions of the test when they are in fact valid.

Of course, we also run the risk of another kind of mistake called a type II error.

A type II error occurs when we fail to reject the assumptions of a test when they have, in fact, been violated.

A test which leads to few type II errors is a high power or sensitive test.
A test which leads to many type II errors is a low power or conservative test.

Clearly, it is better to use a high power test wherever possible since this will lead to a smaller number of type II errors (for a given level of type I errors). In general, tests for normally distributed scores are the most powerful and tests for nominal scores the least powerful. It should be noted, however, that tests for rankable scores are sometimes *almost* as powerful as the most powerful test available. Siegel (1956) discusses the power of many individual tests for rankable and nominal scores. The power of any test always increases with its sample size.

It is unfortunately true that high power tests are typically more effortful than low power tests. This has led to the common strategy of using low effort, low power tests as a first choice. If this test fails to achieve significance then a higher power (and therefore higher effort) test can be used. Whether or not you adopt this procedure is a matter of taste. Many prefer to use the test with the highest possible power on all occasions.

4. Tests for one variable

4.1 General

Single variable tests are essentially tests of distribution, often called 'goodness of fit' tests. The question will often specify a distribution for a population of scores. Our results constitute a sample and we must decide whether our sample could have been drawn from a population with the specified distribution. The test used will depend on the type of score involved. Table 3.3 summarizes the decision procedure.

Nominal scores.　Two tests are available, the binomial and the chi square test. The binomial test should be used in the special case where there are only two nominal categories. This test is concerned with situations where we *expect* the parent population to contain *equal* numbers of scores in the two categories.

　The *chi square* test can handle more than two nominal categories and it can deal with situations where we do not expect the group frequencies to be equal. It has historically proved itself to be very flexible and is very popular.

Rankable scores.　The Kolmogorov–Smirnov (K–S) test applies here. It can be used either for rankable scores or rankable categories. The test is more sensitive when the scores are not categorized. The K–S test is relatively new and unfamiliar. Before its arrival, its job was largely done by the chi square test. This may lead to some confusion since older texts recommend the chi square test for goodness of fit applications even when the scores are rankable. In general, the K–S test should be used for preference when the distribution of scores can be converted to a meaningful cumulative distribution.

4.2 Binomial test

VARIABLE: NOMINAL: 2 CATEGORIES (a, b)
DATA SUMMARY TABLE:

	Categories	
a	b	$a + b$
A	B	N

Notes
A and B are frequencies, i.e., the *number* of scores in categories a and b.
N is the total number of scores; $N = A + B$.

Procedure
1. Take the *smaller* value of A and B and consult table 4.2. Your value must be *equal to or smaller* than the value given in the table to be significant.
   A 1-TAIL test is appropriate when a correct and reasoned prediction was made concerning which value (A or B) would be smaller. Otherwise, use a 2-TAIL test.
2. For *large sample sizes*, compute:

$$Z = \frac{A - B}{\sqrt{(A + B)}} \text{ or } \frac{B - A}{\sqrt{(A + B)}} \quad (4.2.1)$$

Use the formula which yields a positive result. Z is normally distributed and may be assessed with reference to table 2.7.2. Some critical values are, however, given below.

2-TAIL	10%	5%	2%	0·2%
1-TAIL	5%	2·5%	1%	0·1%
$Z \geq$	1·64	1·96	2·33	3·10

Inferences. Table 4.2 describes the situation where the following assumptions are met:

  (i) The sample is drawn at random.
 (ii) The parent population, from which the sample is drawn, contains only two types of score (a and b).
(iii) In the parent population, a and b scores are equally common.

A significantly small value of A (or B) indicates that one of these assumptions is not true. Good experimental procedure should guarantee assumption (i) and the test should not be used when assumption (ii) is not valid. This leaves the implication that assumption (iii) is not true and that our sample is taken from a population where scores of type a and b are not equally common.

▶ Example.   The reasoning test which was given to the students before the beginning of the course normally produces a median score of 50 points when given to college level students. We wish to test whether our group is unusually intelligent or stupid. For a quick test, we select 18 students at random and examine their scores. Do the results show that our group is in any way different from normal college students?

VARIABLE: REASONING ABILITY; NOMINAL, 2 CATEGORIES
        (a = above average, b = below average)
DATA SUMMARY TABLE:

	Categories		
a	b	$a + b$	
7	11	18	$N = 18$

Procedure

1. $A = 7, B = 11$ A is the smaller.

    Consult table 4.2. $N = 18$, $A = 7$.

    Our value of A (7) is not equal to or smaller than the critical value for the 10 per cent significance level ($A = 6$) on a 2-TAIL test.

The 2-TAIL test is appropriate because no prediction had been made whether our students are of above or below average ability.

Conclusions.   Our results were insignificant. This is compatible with the idea that there are as many above as below average students in the total group. Our failure to achieve significance, however, may be due to the small sample size available.

Table 4.2   Critical values of A (or B) in the binomial test.

2-TAIL	10%	5%	2%	0.2%
1-TAIL	5%	2.5%	1%	0.1%
N				
5 A or B $\leq$	0			
6	0			
7	0	0	0	
8	1	0	0	
9	1	1	0	
10	1	1	0	0
11	2	1	1	0
12	2	2	1	0
13	3	2	1	0
14	3	2	2	1
15	3	3	2	1
16	4	3	2	1
17	4	4	3	1
18	5	4	3	2
19	5	4	4	2
20	5	5	4	2
21	6	5	4	3
22	6	5	5	3
23	7	6	5	3
24	7	6	5	4
25	7	7	6	4
26	8	7	6	4
27	8	7	7	5
28	9	8	7	5
29	9	8	7	5
30	10	9	8	6
31	10	9	8	6
32	10	9	8	6
33	11	10	9	7
34	11	10	9	7
35	12	11	10	8
36	12	11	10	8
37	13	12	10	8
38	13	12	11	9
39	13	12	11	9
40	14	13	12	9
41	14	13	12	10
42	15	14	13	10
43	15	14	13	11
44	16	15	13	11
45	16	15	14	11
46	16	15	14	12
47	17	16	15	12
48	17	16	15	12
49	18	17	15	13
50	18	17	16	13

4.3 Chi square test

VARIABLE: NOMINAL: 2 or MORE CATEGORIES (I, II,..., k)
DATA SUMMARY TABLE:

	Categories				
	I	II	...	k	Total
Observed	O_I	O_{II}	...	O_k	N
Expected	E_I	E_{II}	...	E_k	N

Notes
k is the number of categories.
N is the number of scores in the sample.
$O_I ... O_k$ are observed frequencies, i.e., the number of scores in the *sample* which belong to a given category.
$E_I ... E_k$ are expected frequencies, i.e., the number of scores in the sample we expect to belong to each given category.
$\Sigma O = \Sigma E = N$, the total of observed frequencies *must* equal the total of expected frequencies.

Procedure
1. Calculate the expected frequencies from a knowledge of the hypothesized parent population. These expected frequencies need not be equal but they *must* sum to the sample size (N).
2. For each category, calculate

$$\frac{(O - E)^2}{E} \tag{4.3.1}$$

3. Calculate χ^2 by summing these values

$$\chi^2 = \sum \frac{(O - E)^2}{E} \tag{4.3.2}$$

alternatively calculate

$$\chi^2 = \sum \frac{O^2}{E} - N \tag{4.3.3}$$

which is much quicker on a calculating machine.
4. χ^2 has $k - 1$ degrees of freedom.
5. When the sample is of adequate size χ^2 is approximately distributed as chi square and its significance may be assessed by reference to table 2.7.4. Some critical values are given below.

Number of groups	Degrees of freedom	Significance level 5%	2.5%	1%	0.1%
2	1	$\chi^2 \geq 3.9$	5.0	6.6	10.8
3	2	6.0	7.4	9.2	13.8
4	3	7.8	9.3	11.3	16.3
5	4	9.5	11.1	13.3	18.5
6	5	11.1	12.8	15.1	20.5

1-TAIL and 2-TAIL test considerations do not apply here.

Inferences. The approximation to chi square applies when the following assumptions are met:

(i) The sample is drawn at random.
(ii) The sample is of adequate size.
(iii) The expected frequencies (E_i) accurately reflect the relative frequencies of scores in the population from which the sample was drawn.

A significantly large value of χ^2 indicates that one of these assumptions is untrue. Good experimental procedure should guarantee assumption (i). As the sample size decreases, assumption (ii) becomes less valid. As a general rule, the test should not be used if *any* of the expected frequencies is less than 5. This leaves the implication that the expected frequencies do not reflect the population distribution from which our sample was drawn.

▶ Example. The statistics course is open to students from many departments. We wish to know whether this year's enrolment is similar to previous years' in terms of its distribution across departments.

	Botany	Zoology	Psychology	Sociology	Education	Economy	Total
This year	23	41	58	88	30	64	304
Previous years	9%	12%	20%	21%	11%	27%	100%

1. We must adjust the percentage figures for previous years before they can be treated as expected frequencies. We can make this adjustment by multiplying each of last year's figures by

$$\frac{304}{100} = 3.04$$

Expected frequencies	27·36	36·48	60·80	63·84	33·44	82·08	Total 304

2. Using the quick method calculate

$$\chi^2 = \sum \frac{O^2}{E} - N$$

$$= \frac{23^2}{27\cdot36} + \frac{41^2}{36\cdot48} + \frac{58^2}{60\cdot80} + \frac{88^2}{63\cdot84} + \frac{30^2}{33\cdot44} + \frac{64^2}{82\cdot08} - 304$$

$$= 318\cdot86 - 304 = 14\cdot86$$

3. χ^2 has $k - 1 = 6 - 1 = 5\ df$.

4. Our value of χ^2 is significant at better than the 2·5 per cent level. It seems likely that the observed differences are not due simply to random fluctuations but to other effects.

4.4 K–S test for ranked categories (Kolmogorov–Smirnov)

VARIABLE: RANKABLE CATEGORIES $(C_\mathrm{I} > C_\mathrm{II} > \ldots > C_k)$
DATA SUMMARY TABLE:

category	C_I	C_II	$\ldots$	C_k
observed frequency	f_I	f_II	$\ldots$	f_k
cumulative frequency	F_I	F_II	$\ldots$	N
cumulative proportion	p_I	p_II	$\ldots$	$1{\cdot}00$
expected cumulative proportion	P_I	P_II	$\ldots$	$1{\cdot}00$
difference	d_I	d_II	$\ldots$	

Notes
$f_\mathrm{I} \ldots f_k$ are frequencies, i.e., f_k is the number of scores in category k.
N is the total number of scores.

Procedure
1. Change the data summary table so that the frequencies are replaced by *cumulative proportions*. This can be achieved in two steps.
   (a) Form cumulative frequencies by adding each value to all scores to the left.

$$(f_\mathrm{I}), (f_\mathrm{I} + f_\mathrm{II}), (f_\mathrm{I} + f_\mathrm{II} + f_\mathrm{III}), \ldots, (f_\mathrm{I} + f_\mathrm{II} + f_\mathrm{III} + \ldots + f_k)$$

Note that the first value on the left stays the same and the last value on the right will always be N (because it is the sum of all the frequencies).
   (b) Divide all these cumulative frequencies by N.
2. Add the *expected* cumulative proportions, immediately below the observed cumulative proportions.
3. Below this add the value of the difference, d, between the observed and expected frequencies.
4. Choose the largest value (D) which may be positive or negative and consult table 4.4. Your value of D (ignore its sign) must be equal to or larger than the value given in the table to be significant. Your test will normally be a 2-TAIL test except under special circumstances (see Goodman, 1954).
5. For large samples, calculate

$$K = D\sqrt{N} \tag{4.4}$$

To be significant, your value of K must be greater than that shown in this table.

2-TAIL	10%	5%	2%	0·2%
1-TAIL	5%	2·5%	1%	0·1%
$K \geq$	1·22	1·36	1·51	1·86

Inferences. Table 4.4 refers to the situation where the following assumptions are true:

(i) The sample was drawn from the population described by the expected cumulative proportions.
(ii) The sample was drawn at random.
(iii) The categories can be meaningfully ordered.

Significantly large value of D may indicate that one of these assumptions is untrue. Good experimental procedure should guarantee (ii) while the test should not be used if (iii) is not true. This leaves the possibility that assumption (i) is untrue, i.e., that the sample was *not* drawn from the population described by the expected cumulative proportions.

Table 4.4 Critical values of *D* in the K–S test for one variable.

2-TAIL (1-TAIL)	10% (5%)	5% (2·5%)	2% (1%)	0·2% (0·1%)
Sample size *N*				
1	$D \geq 0{\cdot}95$	0·975	0·99	0·999
2	0·78	0·84	0·90	0·97
3	0·64	0·71	0·78	0·90
4	0·57	0·62	0·69	0·82
5	0·51	0·56	0·63	0·75
6	0·47	0·52	0·58	0·70
7	0·44	0·48	0·54	0·65
8	0·41	0·45	0·51	0·61
9	0·39	0·43	0·48	0·58
10	0·37	0·41	0·46	0·56
11	0·35	0·39	0·44	0·53
12	0·34	0·38	0·42	0·51
13	0·33	0·36	0·40	0·49
14	0·31	0·35	0·39	0·48
15	0·30	0·34	0·38	0·46
16	0·29	0·33	0·37	0·45
17	0·29	0·32	0·36	0·43
18	0·28	0·31	0·35	0·42
19	0·27	0·30	0·34	0·41
20	0·26	0·29	0·33	0·40
21	0·26	0·29	0·32	0·39
22	0·25	0·28	0·31	0·38
23	0·25	0·27	0·31	0·38
24	0·24	0·27	0·30	0·37
25	0·24	0·26	0·30	0·36
26	0·23	0·26	0·29	0·35
27	0·23	0·25	0·28	0·34
28	0·22	0·25	0·28	0·34
29	0·22	0·24	0·27	0·33
30	0·22	0·24	0·27	0·33

▶ Example. We wish to know if reasoning ability as measured by our test is normally distributed. We take a random sample of 30 scores:

44, 47, 48, 50, 51, 51, 53, 54, 54, 55, 55, 57, 59, 59, 59, 62, 63, 63, 63, 63, 64, 65, 69, 70, 72, 74, 77, 79, 81, 87.

1. Estimate the mean and standard deviation using formulae (2.5.1) and (2.5.2d)

$$\Sigma X = 1848, \qquad \Sigma X^2 = 117\ 196 \qquad n = 30$$

$$\hat{\mu} = \frac{\Sigma X}{n} \qquad = \frac{1848}{30} = 61 \cdot 6 \tag{2.5.1}$$

$$\hat{\sigma} = \sqrt{\left(\frac{\Sigma X^2 - (\Sigma X)^2/n}{n-1}\right)} \tag{2.5.2d}$$

$$= \sqrt{\left(\frac{117\ 196 - 1848^2/30}{29}\right)} = 10 \cdot 76$$

Convert the scores to Z (standard) scores using formula (2.6.1):

$-1 \cdot 6, -1 \cdot 4, -1 \cdot 3, -1 \cdot 1, -1 \cdot 0, -1 \cdot 0, -0 \cdot 8, -0 \cdot 7, -0 \cdot 7, -0 \cdot 6, -0 \cdot 6, -0 \cdot 4, -0 \cdot 2, -0 \cdot 2, -0 \cdot 2,$
$+0 \cdot 0, +0 \cdot 1, +0 \cdot 1, +0 \cdot 1, +0 \cdot 1, +0 \cdot 2, +0 \cdot 3, +0 \cdot 7, +0 \cdot 8, +1 \cdot 0, +1 \cdot 2, +1 \cdot 4, +1 \cdot 6, +1 \cdot 8, +2 \cdot 4,$

Draw up a frequency table for the data:

Z score	Less than −1·00	−1·0 to −0·49	−0·5 to −0·01	0·0 to 0·5	0·51 to 1·0	More than 1·00
	6	5	4	7	2	6

Give cumulative frequencies and proportions:

Cumulative frequency	6	11	15	22	24	30
Cumulative proportion	0·2	0·37	0·5	0·73	0·8	1·00

2. We now need values for expected frequencies for a normal distribution, and find these directly from table 2.7.2. The cumulative proportions can be found in column 1 when Z is negative and column 2 when Z is positive. Divide percentage values by 100 to obtain proportions.

Z	−1·00	−0·49	0·0	0·5	1·00	>1·00
Expected	0·16	0·31	0·5	0·69	0·84	1·00

3. Compare observed and expected values:

Observed	0·2	0·37	0·5	0·73	0·8	1·0
Expected	0·16	0·31	0·5	0·69	0·84	1·0
d	0·04	0·06	0·0	0·04	0·04	0·0

4. $D = 0 \cdot 06$ (D is maximum value of d): $N = 30$.
Consulting table 4.4 we find that D must be at least 0·24 to be significant at the 5 per cent level. There is no evidence then to contradict the hypothesis that our sample is drawn from a normally distributed population.

4.5 K–S test for rankable scores (Kolmogorov–Smirnov)

VARIABLE: RANKABLE SCORES (X_i)
DATA SUMMARY TABLE

Score	X_1	X_2	X_3	X_4	...	X_N
Observed cumulative proportion	$1/N$	$2/N$	$3/N$	$4/N$	...	1·00
Expected cumulative proportion	P_1	P_2	P_3	P_4	...	1·00
Difference	d_1	d_2	d_3	d_4	...	0·00

Notes

X_1, X_2, etc., are rankable scores set out in ascending order.
N is the number of scores.
P_1, P_2, etc., are the expected cumulative proportions.
   These values are derived from a knowledge of the distribution which the sample is supposed to be drawn from.

Procedure

1. The cumulative proportions can be found simply by dividing the *rank* of each score by N.
2. The *expected* cumulative proportions depend on the population in question.
3. Compute the d values which are the differences between the observed and expected cumulative proportions.
4. Find D, the largest value of d *irrespective of sign*.
5. Refer D to table 4.4. Your value of D must be larger than the value in the table to be significant. Your test will normally be a 2-TAIL test except under special circumstances (see Goodman, 1954).
6. For large samples, compute:

$$K = D \sqrt{N} \tag{4.5}$$

To be significant, K must be greater than the values given in the table below:

2-TAIL	10%	5%	2%	0·2%
1-TAIL	5%	2·5%	1%	0·1%
$K \geq$	1·22	1·36	1·51	1·86

Inferences.   Table 4.4 refers to the situation where the following assumptions are true:

 (i) The sample was drawn from the population described by the expected cumulative proportions.
(ii) The sample was drawn at random.

   Significantly large values of D may indicate that at least one of these two assumptions may be false. Good experimental procedure should guarantee assumption (ii). This leaves the possibility that assumption (i) is not true, i.e., that the sample was not drawn from this population.

▶ Example.   For this example, we can use the same data as the example in section 4.4. We have a sample of 30 reasoning test scores and we wish to decide whether they have been drawn from a normally distributed population. These scores have already been converted to Z scores. We can find the expected cumulative proportions from table 2.7.3 (col. 1 for Z negative and col. 2 for Z positive).

1, 2, 3.

	Z score	Cumulative proportion	Expected	d
1	−1·6	0·03	0·05	0·02
2	−1·4	0·07	0·08	0·01
3	−1·3	0·10	0·10	0·00
4	−1·1	0·13	0·14	0·01
5	−1·0	0·17	0·16	0·01
6	−1·0	0·20	0·16	0·04
7	−0·8	0·23	0·21	0·02
8	−0·7	0·27	0·24	0·03
9	−0·7	0·30	0·24	0·06
10	−0·6	0·33	0·27	0·06
11	−0·6	0·37	0·27	0·10
12	−0·4	0·40	0·34	0·06
13	−0·2	0·43	0·42	0·01
14	−0·2	0·47	0·42	0·05
15	−0·2	0·50	0·42	0·08
16	0·0	0·53	0·50	0·03
17	0·1	0·57	0·54	0·03
18	0·1	0·60	0·54	0·06
19	0·1	0·63	0·54	0·09
20	0·1	0·67	0·54	0·13
21	0·2	0·70	0·58	0·12
22	0·3	0·73	0·62	0·11
23	0·7	0·77	0·76	0·01
24	0·8	0·80	0·79	0·01
25	1·0	0·83	0·80	0·03
26	1·2	0·87	0·88	0·01
27	1·4	0·90	0·92	0·02
28	1·6	0·93	0·95	0·02
29	1·8	0·97	0·96	0·01
30	2·4	1·00	1·00	—

4. The largest value of d is 0·13:

$$D = 0.13$$

5. Consulting table 4.4 we find that D must be at least 0·24 to be significant at the 5 per cent level. Our value is therefore not significant. There is no evidence, then, to contradict the hypothesis that our sample is drawn from a normally distributed population.

5. Tests for two variables (Unrelated scores)

5.1 General

The tests to be described in this section are all tests of association between two variables. They ask whether knowledge of a score on one variable will help us estimate a score on the second variable. For example, if we know a person's sex, does it help us predict their success in an examination? We are asking whether sex matters or whether one's age matters when it comes to examinations. Very commonly, researchers tackle these problems by establishing two groups (males/females, old/young people) and seeing whether the groups differ in their average examination mark. The age problem could also be tackled by having three age groups (young, medium and old). Here we might expect a systematic change across the groups. In yet another design we may simply note a student's age in years and compare it with his examination mark. In this case, we do not have any groups but the situation still concerns the association of two variables. We will need a different test for each experiment but the results of the test will constitute answers to the same problem.

$(A + B)$, A, B. The body of the table gives critical values of C. Your C value must be *equal to or less* than up at random and, where groups are used, the scores in one group are not related to those in another group. The two variables are classified according to the scale of their score. The three types of scale are discussed in section 3.4. They are (a) nominal, (b) rankable, and (c) continuous and normally distributed. Tests using scores on a nominal scale are divided into two kinds, those which deal with (a) two nominal categories, e.g., t test and (b) more than two categories. Categories can be seen to be nominal when they cannot be ranked in any meaningful way, e.g., red, green and yellow peas, English, French and Geography departments. When there are only two groups they are automatically nominal since it is irrelevant how they are ordered, e.g., young/old, male/female.

5.2 Fisher's (2 × 2) exact test

SCORES: ALL INDEPENDENT
VARIABLE: (a) NOMINAL 2 CATEGORIES (GROUPS I, II)
            (b) NOMINAL 2 CATEGORIES (X, Y)
DATA SUMMARY TABLE:

		Category		
		X	Y	Totals
Group I		A	B	$(A + B)$
Group II		C	D	$(C + D)$
		$(A + C)$	$(B + D)$	N

Notes

A, B, C and D are *frequencies*,

e.g. A is the number of scores in group I with 'score' X.

N is the total number of scores.

The table is arranged so that $(A + B)$ is *greater* than $(C + D)$ and A is greater than B. These rearrangements are necessary to allow table 5.2 to be consulted with ease.

Procedure.   Once the summary table has been arranged correctly, no further calculations are necessary and table 5.2 may be consulted directly. Find the appropriate point on the table for your values of N, $(A + B)$, A, B. The body of the table gives critical values of C. Your C value must be *equal to or less* than the values given in order to be statistically significant. Some special cases occur where significance can be achieved with *large values of C*. These do not normally occur when N is small. If in doubt, assess significance using formula (5.2) below.

If no predictions have been made about the results, then the 2-TAIL significance levels are appropriate. If you have predicted the general pattern of scores, then 1-TAIL levels can be used.

For *large samples* when the tables cannot be used, compute Z

$$Z = \frac{|AD - BC| - 0.5N}{\sqrt{\left(\frac{(A + B)(A + C)(B + D)(C + D)}{(N - 1)}\right)}} \tag{5.2}$$

The modulus signs ($\|$) cause us to ignore the sign of the difference before subtracting $0.5N$.

Z is normally distributed and can be assessed by reference to table 2.7.2. Some critical values have been abstracted and are given below:

	10%	5.0%	2%	0.2%
2-TAIL	10%	5.0%	2%	0.2%
1-TAIL	5%	2.5%	1%	0.1%
$Z \geq$	1.64	1.96	2.33	3.10

Inferences. It is helpful to regard the summary table as resulting from an experiment where two samples, groups I and II, were drawn from populations where the only two possible scores are the nominal categories X and Y. Table 5.2 specifies how frequently certain small values of C occur when the following assumptions are met:

(i) The populations, from which our two samples were drawn, contain the same relative frequencies of X and Y scores.
(ii) The two samples were made independently of each other.
(iii) The samples were randomly drawn.

Significantly small values of C indicate that at least one of these assumptions is not true. Proper experimental procedure should guarantee assumptions (ii) and (iii). This leaves the implication that the two samples were drawn from populations where the relative frequencies of X and Y scores are different.

▶ Example. On arriving at university all students about to begin a one year statistics course were assessed with an 'abstract reasoning' test. Two groups of students were selected for further study. One group was a random sample of 14 high scorers while the other group was a random sample of 12 low scorers. An examination at the end of the course rated each student as satisfactory or unsatisfactory. The results for our 26 subjects are given below. Do high scorers on the abstract reasoning test perform any better than low scorers in the examination?

SCORES: INDEPENDENT
VARIABLES: (a) ABSTRACT REASONING; NOMINAL 2 CATEGORIES
(b) EXAMINATION RESULTS; NOMINAL 2 CATEGORIES
DATA SUMMARY TABLE:

	Examination results		
	Unsatisfactory	Satisfactory	Totals
Low	8	4	12
High	2	12	14
			26

Reasoning test score

1. Reorganize the categories so that the higher row total is at the top and so that the higher frequency in the top row is on the left.

	Satisfactory	Unsatisfactory	Totals
High	$A = 12$	$B = 2$	14
Low	$C = 4$	$D = 8$	12
			26

2. Consult table 5.2, where $N = 26$, $(A + B) = 14$, $A = 12$, $B = 2$, $C = 4$. We see that our value of C is equal to the critical value at the 2·5 per cent level of significance (1-TAIL). We use a 1-TAIL test because the better performance of the high scorers on the abstract reasoning test was expected.

Conclusion. The high scorers on the abstract reasoning test have a significantly higher proportion of satisfactory performances in the statistics examination.

Table 5.2   Critical values of C in Fisher's 2×2 exact test.

N	A + B	A	B	2-TAIL 10% / 1-TAIL 5%	5% / 2.5%	2% / 1%	0.2% / 0.1%
8	4	4	0	$C \leq 0$	0		
8	5	5	5	0	0		
8	6	6	0	0	0		
9	5	4	1	0			
9	5	5	0	0	0	0	
9	6	5	1	0			
9	6	6	0	0	0		
9	7	7	0	0			
10	5	4	1	0	0		
10	5	5	0	1	0	0	
10	6	5	1	0	0		
10	6	6	0	1	0	0	
10	7	6	1	0			
10	7	7	0	0	0	0	
10	8	8	0	0	0		
11	6	4	2	0			
11	6	5	1	0	0		
11	6	6	0	1	1	0	
11	7	5	2	0			
11	7	6	1	0			
11	7	7	0	1	0	0	
11	8	7	1	0	0		
11	8	8	0	0	0	0	
11	9	9	0	0	0		
12	6	4	2	0			
12	6	5	1	1	0	0	
12	6	6	0	2	1	1	
12	7	5	2	0			
12	7	6	1	0	0	0	
12	7	7	0	1	1	0	
12	8	6	2	0			
12	8	7	1	0	0		
12	8	8	0	1	1	0	
12	9	7	2	0			
12	9	8	1	0	0		
12	9	9	0	1	0	0	
12	10	9	1	0			
12	10	10	0	0	0		
13	7	4	3	0			
13	7	5	2	0	0		
13	7	6	1	1	0	0	
13	7	7	0	2	1	1	0
13	8	5	3	0			
13	8	6	2	0	0		
13	8	7	1	1	0	0	
13	8	8	0	2	1	1	0
13	9	6	3	0			
13	9	7	2	0	0		
13	9	8	1	0	0	0	
13	9	9	0	1	1	0	
13	10	8	2	0			
13	10	9	1	0	0		
13	10	10	0	1	0	0	
13	11	10	1	0			
13	11	11	0	0	0		

Table 5.2—continued.

N	A + B	A	B	2-TAIL 10% / 1-TAIL 5%	5% / 2·5%	2% / 1%	0·2% / 0·1%
14	7	4	3	C ≤ 0			
14	7	5	2	0	0		
14	7	6	1	1	1	0	
14	7	7	0	3	2	1	0
14	8	5	3	0			
14	8	6	2	0	0		
14	8	7	1	1	1	0	
14	8	8	0	2	2	1	0
14	9	6	3	0			
14	9	7	2	0	0		
14	9	8	1	1	0	0	
14	9	9	0	2	1	1	0
14	10	7	3	0			
14	10	8	2	0	0		
14	10	9	1	1	0	0	
14	10	10	0	1	1	0	0
14	11	9	2	0			
14	11	10	1	0	0		
14	11	11	0	1	0	0	
14	12	11	1	0			
14	12	12	0	0	0		
15	8	5	3	0	0		
15	8	6	2	1	0	0	
15	8	7	1	2	1	1	
15	8	8	0	3	2	2	0
15	9	5	4	0			
15	9	6	3	0	0		
15	9	7	2	1	0	0	
15	9	8	1	1	1	0	
15	9	9	0	2	2	1	0
15	10	6	4	0			
15	10	7	3	0	0		
15	10	8	2	0	0	0	
15	10	9	1	1	1	0	
15	10	10	0	2	1	1	0
15	11	8	3	0			
15	11	9	2	0	0		
15	11	10	1	1	0	0	
15	11	11	0	1	1	1	0
15	12	9	3	0			
15	12	10	2	0	0		
15	12	11	1	0	0	0	
15	12	12	0	1	0	0	
15	13	12	1	0			
15	13	13	0	0	0		
16	8	4	4	0			
16	8	5	3	0	0		
16	8	6	2	1	1	0	
16	8	7	1	2	1	1	0
16	8	8	0	3	3	2	1
16	9	5	4	0			
16	9	6	3	0	0		
16	9	7	2	1	1	0	
16	9	8	1	2	1	1	0
16	9	9	0	3	2	2	1
16	10	6	4	0			
16	10	7	3	0	0		

Table 5.2 Critical values of C in Fisher's 2 × 2 exact test—continued.

N	A + B	A	B	2-TAIL 10% 1-TAIL 5%	5% 2·5%	2% 1%	0·2% 0·1%
16	10	8	2	C ≤ 1	0	0	
16	10	9	1	2	1	1	0
16	10	10	0	3	2	2	0
16	11	7	4	0			
16	11	8	3	0	0		
16	11	9	2	1	0	0	
16	11	10	1	1	1	0	
16	11	11	0	2	2	1	0
16	12	8	4	0			
16	12	9	3	0	0		
16	12	10	2	0	0	0	
16	12	11	1	1	0	0	
16	12	12	0	1	1	1	0
16	13	10	3	0			
16	13	11	2	0	0		
16	13	12	1	0	0	0	
16	13	13	0	1	0	0	
16	14	13	1	0	0		
16	14	14	0	0	0	0	
17	9	5	4	0	0		
17	9	6	3	1	0	0	
17	9	7	2	1	1	0	
17	9	8	1	2	2	1	0
17	9	9	0	4	3	2	1
17	10	5	5	0			
17	10	6	4	0	0		
17	10	7	3	1	0	0	
17	10	8	2	1	1	0	
17	10	9	1	2	2	1	0
17	10	10	0	3	3	2	1
17	11	6	5	0			
17	11	7	4	0	0		
17	11	8	3	1	0	0	
17	11	9	2	1	1	0	
17	11	10	1	2	1	1	0
17	11	11	0	3	2	2	0
17	12	7	5	0			
17	12	8	4	0	0		
17	12	9	3	0	0	0	
17	12	10	2	1	0	0	
17	12	11	1	1	1	0	0
17	12	12	0	2	2	1	0
17	13	9	4	0			
17	13	10	3	0	0		
17	13	11	2	0	0	0	
17	13	12	1	1	1	0	
17	13	13	0	1	1	1	0
17	14	11	3	0			
17	14	12	2	0	0		
17	14	13	1	0	0	0	
17	14	14	0	1	1	0	
17	15	13	2	0			
17	15	14	1	0	0		
17	15	15	0	0	0	0	
18	9	5	4	0	0		
18	9	6	3	1	0	0	
18	9	7	2	2	1	1	
18	9	8	1	3	2	2	0
18	9	9	0	4	4	3	1
18	10	5	5	0			
18	10	6	4	0	0		
18	10	7	3	1	0	0	
18	10	8	2	2	1	1	
18	10	9	1	3	2	1	0
18	10	10	0	4	3	3	1
18	11	6	5	0	0		
18	11	7	4	0	0		
18	11	8	3	1	0	0	
18	11	9	2	2	1	1	
18	11	10	1	2	2	1	0
18	11	11	0	3	3	2	1
18	12	6	6	0			

Table 5.2—continued.

N	A + B	A	B	2-TAIL 10% / 1-TAIL 5%	5% / 2·5%	2% / 1%	0·2% / 0·1%
18	12	7	5	C ≤ 0	0		
18	12	8	4	0	0		
18	12	9	3	1	0	0	
18	12	10	2	1	1	0	
18	12	11	1	2	1	1	0
18	12	12	0	3	2	2	1
18	13	8	5	0			
18	13	9	4	0	0		
18	13	10	3	0	0	0	
18	13	11	2	1	1	0	
18	13	12	1	1	1	1	0
18	13	13	0	2	2	1	0
18	14	9	5	0			
18	14	10	4	0	0		
18	14	11	3	0	0		
18	14	12	2	1	0	0	
18	14	13	1	1	1	0	
18	14	14	0	2	1	1	0
18	15	11	4	0			
18	15	12	3	0	0		
18	15	13	2	0	0		
18	15	14	1	0	0	0	
18	15	15	0	1	1	0	
18	16	14	2	0			
18	16	15	1	0	0		
18	16	16	0	0	0	0	
19	10	5	5	0	0		
19	10	6	4	1	0	0	
19	10	7	3	1	1	0	
19	10	8	2	2	2	1	0
19	10	9	1	3	3	2	0
19	10	10	0	5	4	3	2
19	11	6	5	0	0		
19	11	7	4	1	0	0	
19	11	8	3	1	1	0	
19	11	9	2	2	1	1	0
19	11	10	1	3	2	2	0
19	11	11	0	4	4	3	1
19	12	6	6	0			
19	12	7	5	0	0		
19	12	8	4	1	0	0	
19	12	9	3	1	1	0	
19	12	10	2	2	1	1	0
19	12	11	1	3	2	1	0
19	12	12	0	4	3	2	1
19	13	7	6	0			
19	13	8	5	0	0		
19	13	9	4	0	0	0	
19	13	10	3	1	0	0	
19	13	11	2	1	1	0	
19	13	12	1	2	2	1	0
19	13	13	0	3	2	2	1
19	14	8	6	0			
19	14	9	5	0	0		
19	14	10	4	0	0		
19	14	11	3	1	0	0	
19	14	12	2	1	1	0	
19	14	13	1	2	1	1	0
19	14	14	0	2	2	1	0
19	15	10	5	0			
19	15	11	4	0	0		
19	15	12	3	0	0	0	
19	15	13	2	1	0	0	
19	15	14	1	1	1	0	
19	15	15	0	2	1	1	0
19	16	12	4	0			
19	16	13	3	0	0		
19	16	14	2	0	0		
19	16	15	1	0	0	0	
19	16	15	1	1	1	0	
19	17	15	2	0			
19	17	16	1	0	0		
19	17	17	0	0	0	0	

Table 5.2 Critical values of C in Fisher's 2 × 2 exact test—continued.

N	A + B	A	B	2-TAIL 10% 1-TAIL 5%	5% 2·5%	2% 1%	0·2% 0·1%
20	10	5	5	C ≤ 0	0		
20	10	6	4	1	0	0	
20	10	7	3	2	1	0	
20	10	8	2	3	2	1	0
20	10	9	1	4	3	2	1
20	10	10	0	5	5	4	2
20	11	6	5	0	0		
20	11	7	4	1	0	0	
20	11	8	3	2	1	0	
20	11	9	2	2	2	1	0
20	11	10	1	3	3	2	1
20	11	11	0	5	4	3	2
20	12	6	6	0	0		
20	12	7	5	0	0		
20	12	8	4	1	0	0	
20	12	9	3	2	1	0	
20	12	10	2	2	2	1	0
20	12	11	1	3	3	2	1
20	12	12	0	4	4	3	2
20	13	7	6	0	0		
20	13	8	5	0	0		
20	13	9	4	1	0	0	
20	13	10	3	1	1	0	
20	13	11	2	2	1	1	0
20	13	12	1	3	2	2	0
20	13	13	0	4	3	3	1
20	14	7	7	0			
20	14	8	6	0	0		
20	14	9	5	0	0		
20	14	10	4	1	0	0	
20	14	11	3	1	1	0	
20	14	12	2	2	1	1	0
20	14	13	1	2	2	1	0
20	14	14	0	3	3	2	1
20	15	9	6	0			
20	15	10	5	0	0		
20	15	11	4	0	0	0	
20	15	12	3	1	0	0	
20	15	13	2	1	1	0	
20	15	14	1	2	1	1	0
20	15	15	0	2	2	2	0
20	16	10	6	0			
20	16	11	5	0			
20	16	12	4	0	0		
20	16	13	3	0	0	0	
20	16	14	2	1	0	0	
20	16	15	1	1	1	0	
20	16	16	0	2	1	1	0
20	17	12	5	0			
20	17	13	4	0			
20	17	14	3	0	0		
20	17	15	2	0	0	0	
20	17	16	1	1	0	0	
20	17	17	0	1	1	0	0
20	18	16	2	0			
20	18	17	1	0	0		
20	18	18	0	0	0	0	
20	19	19	0	0			
21	11	6	5	1	0	0	
21	11	7	4	1	1	0	
21	11	8	3	2	1	1	0
21	11	9	2	3	2	2	0
21	11	10	1	4	3	3	1
21	11	11	0	5	5	4	2
21	12	6	6	0	0		
21	12	7	5	1	0	0	
21	12	8	4	1	1	0	
21	12	9	3	2	1	1	0
21	12	10	2	3	2	2	0
21	12	11	1	4	3	2	1

Table 5.2—continued.

N	A + B	A	B	2-TAIL 10% / 1-TAIL 5%	5% / 2·5%	2% / 1%	0·2% / 0·1%
21	12	12	0	$C \leq$ 5	4	4	2
21	13	7	6	0	0		
21	13	8	5	1	0	0	
21	13	9	4	1	1	0	
21	13	10	3	2	1	1	0
21	13	11	2	2	2	1	0
21	13	12	1	3	3	2	1
21	13	13	0	4	4	3	2
21	14	7	7	0			
21	14	8	6	0	0		
21	14	9	5	1	0	0	
21	14	10	4	1	1	0	
21	14	11	3	2	1	1	
21	14	12	2	2	2	1	0
21	14	13	1	3	2	2	1
21	14	14	0	4	3	3	1
21	15	8	7	0			
21	15	9	6	0	0		
21	15	10	5	0	0	0	
21	15	11	4	1	0	0	
21	15	12	3	1	1	0	
21	15	13	2	2	1	1	0
21	15	14	1	2	2	1	0
21	15	15	0	3	3	2	1
21	16	9	7	0			
21	16	10	6	0	0		
21	16	11	5	0	0		
21	16	12	4	0	0	0	
21	16	13	3	1	0	0	
21	16	14	2	1	1	0	
21	16	15	1	2	1	1	0
21	16	16	0	2	2	2	1
21	17	11	6	0			
21	17	12	5	0	0		
21	17	13	4	0	0		
21	17	14	3	0	0	0	
21	17	15	2	1	0	0	
21	17	16	1	1	1	0	0
21	17	17	0	2	1	1	0
21	18	13	5	0			
21	18	14	4	0			
21	18	15	3	0	0		
21	18	16	2	0	0	0	
21	18	17	1	1	0	0	
21	18	18	0	1	1	0	0
21	19	16	3	0			
21	19	17	2	0			
21	19	18	1	0	0		
21	19	19	0	0	0	0	
21	20	20	0	0			
22	11	6	5	1	0	0	
22	11	7	4	2	1	0	
22	11	8	3	2	2	1	0
22	11	9	2	3	3	2	0
22	11	10	1	5	4	3	1
22	11	11	0	6	5	5	3
22	12	6	6	0	0		
22	12	7	5	1	0	0	
22	12	8	4	2	1	0	
22	12	9	3	2	2	1	0
22	12	10	2	3	3	2	0
22	12	11	1	4	4	3	1
22	12	12	0	6	5	4	3
22	13	7	6	0	0		
22	13	8	5	1	0	0	
22	13	9	4	2	1	0	
22	13	10	3	2	2	1	0
22	13	11	2	3	2	2	0
22	13	12	1	4	3	3	1

Table 5.2   Critical values of C in Fisher's 2 × 2 exact test—continued.

N	A + B	A	B	2-TAIL 10% 1-TAIL 5%	5% 2·5%	2% 1%	0·2% 0·1%
22	13	13	0	C ≤ 5	5	4	2
22	14	7	7	0	0		
22	14	8	6	0	0	0	
22	14	9	5	1	0	0	
22	14	10	4	1	1	0	
22	14	11	3	2	1	1	0
22	14	12	2	3	2	2	0
22	14	13	1	3	3	2	1
22	14	14	0	5	4	3	2
22	15	8	7	0	0		
22	15	9	6	0	0		
22	15	10	5	1	0	0	
22	15	11	4	1	1	0	
22	15	12	3	2	1	1	0
22	15	13	2	2	2	1	0
22	15	14	1	3	3	2	1
22	15	15	0	4	3	3	2
22	16	8	8	0			
22	16	9	7	0	0		
22	16	10	6	0	0		
22	16	11	5	1	0	0	
22	16	12	4	1	1	Q	
22	16	13	3	1	1	0	
22	16	14	2	2	1	1	0
22	16	15	1	2	2	1	0
22	16	16	0	3	3	2	1
22	17	9	8	0			
22	17	10	7	0			
22	17	11	6	0	0		
22	17	12	5	0	0	0	
22	17	13	4	1	0	0	
22	17	14	3	1	1	0	
22	17	15	2	1	1	1	0
22	17	16	1	2	1	1	0
22	17	17	0	2	2	2	1
22	18	11	7	0			
22	18	12	6	0			
22	18	13	5	0	0		
22	18	14	4	0	0	0	
22	18	15	3	1	0	0	
22	18	16	2	1	0	0	
22	18	17	1	1	1	0	0
22	18	18	0	2	1	1	0
22	19	14	5	0			
22	19	15	4	0	0		
22	19	16	3	0	0		
22	19	17	2	0	0	0	
22	19	18	1	1	0	0	
22	19	19	0	1	1	0	0
22	20	17	3	0			
22	20	18	2	0			
22	20	19	1	0	0		
22	20	20	0	0	0	0	
22	21	21	0	0			
23	12	6	6	0	0	0	
23	12	7	5	1	1	0	
23	12	8	4	2	1	1	
23	12	9	3	3	2	1	0
23	12	10	2	4	3	2	1
23	12	11	1	5	4	3	2
23	12	12	0	6	6	5	3
23	13	7	6	1	0	0	
23	13	8	5	1	1	0	
23	13	9	4	2	1	1	0
23	13	10	3	3	2	1	0

Table 5.2—continued.

N	A + B	A	B	2-TAIL 10% 1-TAIL 5%	5% 2·5%	2% 1%	0·2% 0·1%
23	13	11	2	$C \leq$ 4	3	2	1
23	13	12	1	5	4	3	2
23	13	13	0	6	5	4	3
23	14	7	7	0	0		
23	14	8	6	1	0	0	
23	14	9	5	1	1	0	
23	14	10	4	2	1	1	0
23	14	11	3	2	2	1	0
23	14	12	2	3	3	2	1
23	14	13	1	4	4	3	1
23	14	14	0	5	5	4	3
23	15	8	7	0	0		
23	15	9	6	1	0	0	
23	15	10	5	1	1	0	
23	15	11	4	2	1	1	
23	15	12	3	2	2	1	0
23	15	13	2	3	2	2	0
23	15	14	1	4	3	2	1
23	15	15	0	5	4	4	2
23	16	8	8	0			
23	16	9	7	0	0		
23	16	10	6	1	0	0	
23	16	11	5	1	0	0	
23	16	12	4	1	1	0	
23	16	13	3	2	1	1	0
23	16	14	2	2	2	1	0
23	16	15	1	3	3	2	1
23	16	16	0	4	4	3	2
23	17	9	8	0			
23	17	10	7	0	0		
23	17	11	6	0	0	0	
23	17	12	5	1	0	0	
23	17	13	4	1	1	0	
23	17	14	3	1	1	1	0
23	17	15	2	2	2	1	0
23	17	16	1	3	2	2	0
23	17	17	0	3	3	2	1
23	18	10	8	0			
23	18	11	7	0	0		
23	18	12	6	0	0		
23	18	13	5	0	0	0	
23	18	14	4	1	0	0	
23	18	15	3	1	1	0	
23	18	16	2	1	1	1	0
23	18	17	1	2	2	1	0
23	18	18	0	3	2	2	1
23	19	12	7	0			
23	19	13	6	0	0		
23	19	14	5	0	0		
23	19	15	4	0	0	0	
23	19	16	3	1	0	0	
23	19	17	2	1	1	0	
23	19	18	1	1	1	1	0
23	19	19	0	2	1	1	0
23	20	14	6	0			
23	20	15	5	0			
23	20	16	4	0	0		
23	20	17	3	0	0		
23	20	18	2	0	0	0	
23	20	19	1	1	0	0	
23	20	20	0	1	1	0	0
23	21	18	3	0			
23	21	19	2	0	0		
23	21	20	1	0	0		
23	21	21	0	0	0	0	
23	22	22	0	0			

Table 5.2   Critical values of C in Fisher's 2 × 2 exact test—continued.

N	$A+B$	A	B	2-TAIL 10% 1-TAIL 5%	5% 2·5%	2% 1%	0·2% 0·1%
24	12	6	6	$C \le 1$	0	0	
24	12	7	5	1	1	0	
24	12	8	4	2	2	1	0
24	12	9	3	3	2	2	0
24	12	10	2	4	4	3	1
24	12	11	1	5	5	4	2
24	12	12	0	7	6	5	4
24	13	7	6	1	0	0	
24	13	8	5	2	1	0	
24	13	9	4	2	2	1	0
24	13	10	3	3	2	2	0
24	13	11	2	4	3	3	1
24	13	12	1	5	4	4	2
24	13	13	0	7	6	5	3
24	14	7	7	0	0	0	
24	14	8	6	1	0	0	
24	14	9	5	2	1	0	
24	14	10	4	2	2	1	0
24	14	11	3	3	2	2	0
24	14	12	2	4	3	2	1
24	14	13	1	5	4	3	2
24	14	14	0	6	5	5	3
24	15	8	7	0	0	0	
24	15	9	6	1	0	0	
24	15	10	5	1	1	0	
24	15	11	4	2	2	1	0
24	15	12	3	3	2	1	0
24	15	13	2	3	3	2	1
24	15	14	1	4	4	3	2
24	15	15	0	5	5	4	3
24	16	8	8	0	0		
24	16	9	7	0	0	0	
24	16	10	6	1	0	0	
24	16	11	5	1	1	0	
24	16	12	4	2	1	1	0
24	16	13	3	2	2	1	0
24	16	14	2	3	3	2	1
24	16	15	1	4	3	3	1
24	16	16	0	5	4	4	2
24	17	9	8	0	0		
24	17	10	7	0	0	0	
24	17	11	6	1	0	0	
24	17	12	5	1	1	0	
24	17	13	4	2	1	1	0
24	17	14	3	2	2	1	0
24	17	15	2	3	2	2	0
24	17	16	1	3	3	2	1
24	17	17	0	4	4	3	2
24	18	9	9	0			
24	18	10	8	0	0		
24	18	11	7	0	0		
24	18	12	6	0	0	0	
24	18	13	5	1	0	0	
24	18	14	4	1	1	0	
24	18	15	3	2	1	1	0
24	18	16	2	2	2	1	0
24	18	17	1	3	2	2	1
24	18	18	0	3	3	2	1
24	19	10	9	0			
24	19	11	8	0			
24	19	12	7	0	0		
24	19	13	6	0	0		
24	19	14	5	0	0	0	
24	19	15	4	1	0	0	
24	19	16	3	1	1	0	
24	19	17	2	2	1	1	0
24	19	18	1	2	2	1	0
24	19	19	0	3	2	2	1
24	20	12	8	0			
24	20	13	7	0			
24	20	14	6	0	0		

Table 5.2—continued.

N	A + B	A	B	2-TAIL 10% 1-TAIL 5%	5% 2·5%	2% 1%	0·2% 0·1%
24	20	15	5	C ≤ 0	0		
24	20	16	4	0	0	0	
24	20	17	3	1	0	0	
24	20	18	2	1	1	0	
24	20	19	1	1	1	1	0
24	20	20	0	2	2	1	0
24	21	15	6	0			
24	21	16	5	0			
24	21	17	4	0	0		
24	21	18	3	0	0	0	
24	21	19	2	0	0	0	
24	21	20	1	1	0	0	
24	21	21	0	1	1	0	0
24	22	19	3	0			
24	22	20	2	0	0		
24	22	21	1	0	0		
24	22	22	0	0	0	0	
24	23	23	0	0			
25	13	7	6	1	1	0	
25	13	8	5	2	1	1	
25	13	9	4	3	2	1	0
25	13	10	3	4	3	2	1
25	13	11	2	5	4	3	1
25	13	12	1	6	5	4	2
25	13	13	0	7	7	6	4
25	14	7	7	1	0	0	
25	14	8	6	1	1	0	
25	14	9	5	2	2	1	0
25	14	10	4	3	2	1	0
25	14	11	3	3	3	2	1
25	14	12	2	4	4	3	1
25	14	13	1	5	5	4	2
25	14	14	0	7	6	5	4
25	15	8	7	1	0	0	
25	15	9	6	1	1	0	
25	15	10	5	2	1	1	0
25	15	11	4	2	2	1	0
25	15	12	3	3	3	2	1
25	15	13	2	4	3	3	1
25	15	14	1	5	4	4	2
25	15	15	0	6	6	5	3
25	16	8	8	0	0		
25	16	9	7	1	0	0	
25	16	10	6	1	1	0	
25	16	11	5	2	1	1	0
25	16	12	4	2	2	1	0
25	16	13	3	3	2	2	0
25	16	14	2	4	3	2	1
25	16	15	1	4	4	3	2
25	16	16	0	6	5	4	3
25	17	9	8	0	0		
25	17	10	7	1	0	0	
25	17	11	6	1	1	0	
25	17	12	5	2	1	0	
25	17	13	4	2	1	1	0
25	17	14	3	3	2	1	0
25	17	15	2	3	3	2	1
25	17	16	1	4	3	3	1
25	17	17	0	5	4	4	2
25	18	9	9	0	0		
25	18	10	8	0	0		
25	18	11	7	1	0	0	
25	18	12	6	1	0	0	
25	18	13	5	1	1	0	
25	18	14	4	2	1	1	0
25	18	15	3	2	2	1	0
25	18	16	2	3	2	2	0
25	18	17	1	3	3	2	1
25	18	18	0	4	4	3	2
25	19	10	9	0			

Table 5.2   Critical values of C in Fisher's 2 × 2 exact test—continued.

N	A + B	A	B	2-TAIL 10% 1-TAIL 5%	5% 2·5%	2% 1%	0·2% 0·1%
25	19	11	8	C ≤ 0	0		
25	19	12	7	0	0	0	
25	19	13	6	1	0	0	
25	19	14	5	1	1	0	
25	19	15	4	1	1	0	
25	19	16	3	2	1	1	0
25	19	17	2	2	2	1	0
25	19	18	1	3	2	2	1
25	19	19	0	3	3	3	1
25	20	11	9	0			
25	20	12	8	0	0		
25	20	13	7	0	0		
25	20	14	6	0	0	0	
25	20	15	5	1	0	0	
25	20	16	4	1	1	0	
25	20	17	3	1	1	0	
25	20	18	2	2	1	1	0
25	20	19	1	2	2	1	0
25	20	20	0	3	2	2	1
25	21	13	8	0			
25	21	14	7	0			
25	21	15	6	0	0		
25	21	16	5	0	0	0	
25	21	17	4	0	0	0	
25	21	18	3	1	0	0	
25	21	19	2	1	1	0	
25	21	20	1	1	1	1	0
25	21	21	0	2	2	1	0
25	22	16	6	0			
25	22	17	5	0	0		
25	22	18	4	0	0		
25	22	19	3	0	0	0	
25	22	20	2	0	0	0	
25	22	21	1	1	0	0	
25	22	22	0	1	1	0	0
25	23	19	4	0			
25	23	20	3	0			
25	23	21	2	0	0		
25	23	22	1	0	0	0	
25	23	23	0	0	0	0	
25	24	24	0	0			
26	13	7	6	1	1	0	
26	13	8	5	2	1	1	0
26	13	9	4	3	2	2	0
26	13	10	3	4	3	2	1
26	13	11	2	5	4	3	2
26	13	12	1	6	6	5	3
26	13	13	0	8	7	6	4
26	14	7	7	1	0	0	
26	14	8	6	1	1	0	
26	14	9	5	2	2	1	0
26	14	10	4	3	2	2	0
26	14	11	3	4	3	2	1
26	14	12	2	5	4	3	2
26	14	13	1	6	5	4	3
26	14	14	0	7	7	6	4
26	15	8	7	1	0	0	
26	15	9	6	2	1	0	
26	15	10	5	2	2	1	0
26	15	11	4	3	2	2	0
26	15	12	3	4	3	2	1
26	15	13	2	5	4	3	2
26	15	14	1	6	5	4	3
26	15	15	0	7	6	6	4
26	16	8	8	1	0	0	
26	16	9	7	1	0	0	
26	16	10	6	2	1	0	

Table 5.2—continued.

N	A + B	A	B	2-TAIL 10% 1-TAIL 5%	5% 2·5%	2% 1%	0·2% 0·1%
26	16	11	5	$C \leq$ 2	2	1	0
26	16	12	4	3	2	1	0
26	16	13	3	3	3	2	1
26	16	14	2	4	4	3	1
26	16	15	1	5	5	4	2
26	16	16	0	6	6	5	3
26	17	9	8	1	0	0	
26	17	10	7	1	0	0	
26	17	11	6	1	1	0	
26	17	12	5	2	1	1	0
26	17	13	4	2	2	1	0
26	17	14	3	3	3	2	1
26	17	15	2	4	3	3	1
26	17	16	1	5	4	3	2
26	17	17	0	6	5	4	3
26	18	9	9	0	0		
26	18	10	8	0	0	0	
26	18	11	7	1	0	0	
26	18	12	6	1	1	0	
26	18	13	5	2	1	1	0
26	18	14	4	2	2	1	0
26	18	15	3	3	2	2	0
26	18	16	2	3	3	2	1
26	18	17	1	4	4	3	2
26	18	18	0	5	4	4	3
26	19	10	9	0	0		
26	19	11	8	0	0	0	
26	19	12	7	1	0	0	
26	19	13	6	1	1	0	
26	19	14	5	1	1	0	
26	19	15	4	2	1	1	0
26	19	16	3	2	2	1	0
26	19	17	2	3	2	2	1
26	19	18	1	3	3	2	1
26	19	19	0	4	4	3	2
26	20	10	10	0			
26	20	11	9	0	0		
26	20	12	8	0	0		
26	20	13	7	0	0	0	
26	20	14	6	1	0	0	
26	20	15	5	1	1	0	
26	20	16	4	1	1	0	0
26	20	17	3	2	1	1	0
26	20	18	2	2	2	1	0
26	20	19	1	3	2	2	1
26	20	20	0	3	3	3	1
26	21	11	10	0			
26	21	12	9	0			
26	21	13	8	0	0		
26	21	14	7	0	0		
26	21	15	6	0	0	0	
26	21	16	5	1	0	0	
26	21	17	4	1	1	0	
26	21	18	3	1	1	0	0
26	21	19	2	2	1	1	0
26	21	20	1	2	2	1	0
26	21	21	0	3	2	2	1
26	22	13	9	0			
26	22	14	8	0			
26	22	15	7	0	0		
26	22	16	6	0	0		
26	22	17	5	0	0	0	
26	22	18	4	1	0	0	
26	22	19	3	1	0	0	
26	22	20	2	1	1	0	
26	22	21	1	1	1	1	0
26	22	22	0	2	2	1	0
26	23	16	7	0			

Table 5.2 Critical values of C in Fisher's 2 × 2 exact test—continued.

N	A + B	A	B	2-TAIL 10% 1-TAIL 5%	5% 2·5%	2% 1%	2% 0·1%
26	23	17	6	$C \leq$ 0			
26	23	18	5	0	0		
26	23	19	4	0	0		
26	23	20	3	0	0	0	
26	23	21	2	0	0	0	
26	23	22	1	1	0	0	
26	23	23	0	1	1	1	0
26	24	20	4	0			
26	24	21	3	0			
26	24	22	2	0	0		
26	24	23	1	0	0	0	
26	24	24	0	0	0	0	
26	25	25	0	0			

5.3 Chi square (2 × k) test

SCORES: ALL INDEPENDENT
VARIABLES: (a) NOMINAL 2 CATEGORIES (Groups I, II)
 (b) NOMINAL 2 OR MORE CATEGORIES $(A_1, A_2, \ldots, A_k)$
DATA SUMMARY TABLE:

	Category				Row totals
	A_1	A_2	$\ldots$	A_k	
Group I	$O_{1,1}$	$O_{1,2}$	$\ldots$	$O_{1,k}$	R_1
Group II	$O_{2,1}$	$O_{2,2}$	$\ldots$	$O_{2,k}$	R_2
Column totals	C_1	C_2	$\ldots$	C_k	N

Notes

O are the *observed frequencies*.
R_1 and R_2 are the row totals.
$C_1, C_2, \ldots, C_k$ are the column totals.
N is the grand total.

Procedure

1. Compute the *expected frequencies* $(E_{i,j})$ for each cell: multiply the row and column total for that cell, then divide by N.

$$E_{i,j} = \frac{R_i \times C_j}{N} \qquad (5.3.1)$$

	A_1	A_2	$\ldots$	A_k	
I	$E_{1,1} = \dfrac{R_1 \times C_1}{N}$	$E_{1,2} = \dfrac{R_1 \times C_2}{N}$	$\ldots$	$E_{1,k} = \dfrac{R_1 \times C_k}{N}$	R_1
II	$E_{2,1} = \dfrac{R_2 \times C_1}{N}$	$E_{2,2} = \dfrac{R_2 \times C_2}{N}$	$\ldots$	$E_{2,k} = \dfrac{R_2 \times C_k}{N}$	R_2
	C_1	C_2	$\ldots$	C_k	N

Check that the expected frequencies sum to the same row and column totals as the observed frequencies.

2. *For each cell*, compute:

$$\frac{(O - E)^2}{E} \qquad (5.3.2)$$

	A_1	A_2	$\ldots$	A_k
I	$\dfrac{(O_{1,1} - E_{1,1})^2}{E_{1,1}}$	$\dfrac{(O_{1,2} - E_{1,2})^2}{E_{1,2}}$	$\ldots$	$\dfrac{(O_{1,k} - E_{1,k})^2}{E_{1,k}}$
II	$\dfrac{(O_{2,1} - E_{2,1})^2}{E_{2,1}}$	$\dfrac{(O_{2,2} - E_{2,2})^2}{E_{2,2}}$	$\ldots$	$\dfrac{(O_{2,k} - E_{2,k})^2}{E_{2,k}}$

3. Find χ^2 which is the sum of all of these values

$$\chi^2 = \sum \frac{(O - E)^2}{E} \tag{5.3.3}$$

χ^2 has $k - 1$ degrees of freedom.

4. Our value of χ^2 may now be assessed for significance by consulting table 2.7.4.

Inferences. Table 2.7.4 specifies how frequently certain values of χ^2 are exceeded when the following assumptions are met:

(i) The populations, from which our two samples are drawn, contain the *same* relative frequencies of scores in the various categories $(A_1, A_2, \dots, A_k)$.

(ii) The two samples were drawn independently.

(iii) The scores within each sample were drawn at random.

(iv) The samples are of adequate size; see below.

Significantly large values of χ^2 indicate that at least one of these assumptions is not true. Proper experimental procedure should guarantee assumption (iii). The test should not be used when assumptions (ii) and (iv) are not met by the data. This leaves the implication that our two samples were drawn from populations with differing relative frequencies of scores in the categories $A_1, A_2, \dots, A_k$.

Sample sizes. In this test, the χ^2 values given in the significance tables are only approximately valid. For large sample sizes, the approximation is excellent. It is advisable *not* to use this test if more than 20 per cent of the expected frequencies $(E_{i,j})$ are less than 5.

▶ Example. Students from two scientific disciplines (physical sciences and life sciences) attend a course on introductory statistics. At the end of the course, they are allowed one of three options: (a) to proceed to a more advanced course, (b) to repeat the same course (usually obligatory) or (c) to attend no more statistics courses. The number of options taken by random samples of social and life science students are summarized in the table below. Is there any evidence to show that the options taken by the students depend upon their particular scientific discipline?

SCORES: ALL INDEPENDENT
VARIABLES: (a) SCIENTIFIC DISCIPLINE; NOMINAL, 2 CATEGORIES
            (b) OPTION TAKEN; NOMINAL, 3 CATEGORIES
DATA SUMMARY TABLE:

		Option			
		(a) Advanced course	(b) Repeat course	(c) No more courses	
Discipline	Life sciences	6	6	28	40
	Physical sciences	17	7	11	35
		23	13	39	75

1. Compute the expected frequencies using formula (5.3.1):

	(a)	(b)	(c)	
Life sciences	12·27	6·93	20·80	40
Physical sciences	10·73	6·07	18·20	35
	23	13	39	75

Then check that the row and column totals agree with those in the data summary table.

2. Compute for each cell:

$$\frac{(O - E)^2}{E}$$

	(a)	(b)	(c)
Life sciences	3·2	0·12	2·49
Physical sciences	3·66	0·14	2·85

3. Compute χ^2 using formula (5.3.3)

$$\chi^2 = \sum \frac{(O - E)^2}{E} = 12·46, \qquad df = 3 - 1 = 2$$

Conclusions. Our χ^2 value is significant at the 1·0 per cent level and we may conclude that the pattern of options is different in the two disciplines.

5.4   K–S (Kolmogorov–Smirnov) two sample test

SCORES: INDEPENDENT
VARIABLES: (a) NOMINAL; 2 GROUPS (I, II)
             (b) RANKED CATEGORIES ($C_1 < C_2 < \ldots < C_k$)
DATA SUMMARY TABLE

(b) Ranked categories

	$C_1 < C_2 < \ldots < C_k$				Sample size
Group I	F	F	$\ldots$	F	n_1
Group II	F	F	$\ldots$	F	n_2

Notes

F refers to the *frequency* of scores in each group/category combination.

n_1, n_2 are sample sizes and are the sum of frequencies in that group.

Procedure

1. Recast the data summary table in terms of cumulative frequencies for each group. Work from left to right.
2. Convert each cumulative frequency to a cumulative proportion (CP) by dividing by the appropriate sample size. If this is done correctly, both figures in the end column will be 1·00.
3. In each column, subtract the CP in the bottom row from the CP in the top row. This value is called d and is shown in the table.

$C_1 < C_2 < \ldots < C_k$

Group I	CP	CP	$\ldots$	1·00
Group II	CP	CP	$\ldots$	1·00
Difference	d	d	$\ldots$	0

4. Find D, the largest value of d (irrespective of sign).
5. Compute K, where

$$K = D \sqrt{\frac{n_1 n_2}{n_1 + n_2}} \tag{5.4}$$

The significance of K may be assessed by reference to table 5.4. Your value of K must be bigger than the value specified in the table to be significant.

A 1-TAIL test is appropriate only if the sign of D (positive or negative) was successfully predicted before the results were available.

Table 5.4.   Critical values of K in the K–S two sample test.

2-TAIL	10%	5%	2%	0·2%
1-TAIL	5%	2·5%	1%	0·1%
$K \geq$	1·22	1·36	1·51	1·86

Inferences. As in sections 4.4 and 4.5 the K–S test is a test of difference in distribution function. The various significance tables describe the situation where the following assumptions are met:

 (i) The two samples are drawn from populations with the same distribution characteristics (mean, variance, skew, range, etc.).
 (ii) The two samples were drawn independently.
(iii) The scores within each sample were drawn at random.
(iv) The ordering of the categories is meaningful and not arbitrary.

A significant result indicates that at least one of the assumptions is not true. The test should not be used if assumptions (ii) and (iv) are not met and proper experimental procedure should guarantee assumption (iii). This leaves the possibility that some characteristic of the distributions is different. For a 1-TAIL test, this usually indicates that the means are different. For a 2-TAIL test, this is not always the case and careful consideration should be given to the possibility that the major difference may involve skew.

▶ Example.   On arrival at university, all students about to begin a one year course in statistics were assessed with an 'abstract reasoning' test. Two groups of students were selected for further study. One group was a random sample of 8 high scorers while the other group was a random sample of 10 low scorers. An examination at the end of the course rated each student as either 'fail', 'poor', 'satisfactory', good' or 'excellent'. The results are given below. Do the results confirm our prediction that high scorers will do well in the examination?

SCORES: INDEPENDENT
VARIABLES: (a) ABSTRACT REASONING; NOMINAL 2 CATEGORIES
            (b) EXAMINATION RESULTS; 5 RANKED CATEGORIES
DATA SUMMARY TABLE:

| | Examination results | | | | | |
Reasoning test score	Fail	Poor	Satisfactory	Good	Excellent	Sample size
High	—	1	1	5	1	8
Low	2	3	4	1	—	10

1. Recast the table in cumulative frequencies

	—	1	2	7	8
	2	5	9	10	10

2. Convert to cumulative *proportions* by dividing by the appropriate sample size.
3. Compute d for each column

	—	0·125	0·250	0·875	1·00
	0·200	0·500	0·900	1·000	1·00
d	−0·200	−0·375	−0·650	−0·125	0

4. The largest value of d is -0.650; therefore D is 0.65.
5. Compute K using formula (5.4)

$$K = 0.65 \sqrt{\frac{8 \times 10}{8 + 10}} = 1.37$$

Using a 1-TAIL test, and consulting table 5.4, we find that our value of K is significant at better than the 2·5 per cent significance level. We used a 1-TAIL test because we predicted that high scores on the reasoning test would produce better examination results.

Conclusion.   We conclude that high scorers on the reasoning tests are more likely to perform better in the examination.

5.5   Mann–Whitney test

SCORES: INDEPENDENT
VARIABLES: (a) NOMINAL; 2 GROUPS (I, II)
                (b) RANKABLE SCORES
DATA SUMMARY TABLE:

					Sample size
Group I	x_1	x_2	x_3 $\cdots$	x_{n_1}	n_1
Group II	y_1	y_2	y_3 $\cdots$	y_{n_2}	n_2

Notes
x_i and y_i are rankable scores.
n_1 is the number of scores in group I.
n_2 is the number of scores in group II.
$n_1 \leq n_2$; group I is never larger than group II.

Procedure
1. Rank all the scores (irrespective of group) from 1 to $n_1 + n_2$.
2. Find the sum of the ranks (R) in group I, the smaller group.
3. Calculate R'

$$R' = n_1(n_1 + n_2 + 1) - R$$

and use the *smaller* value of R' or R when consulting table 5.5.
4. The significance of R (or R') can be assessed directly with reference to table 5.5 for small samples. The value of R (or R') must be *equal to or smaller than* the value in the table to be significant. If the investigator has successfully predicted whether R or R' would be smaller, then a 1-TAIL test is appropriate.

For large samples, we may compute:

$$Z = \frac{n_1(n_1 + n_2 + 1) - 2R}{\sqrt{\left(\dfrac{n_1 n_2 (n_1 + n_2 + 1)}{3} \right)}} \tag{5.5}$$

The significance of Z may be assessed with reference to table 2.7.2 but relevant critical values are given below.

2-TAIL	10%	5%	2%	0·2%
1-TAIL	5%	2·5%	1%	0·1%
$Z \geq$	1·64	1·96	2·33	3·10

The test is unreasonably tedious for large samples where it can be considerably quicker to group the data and use the K–S two sample test instead.

Inferences.   Our two groups are samples from two populations and we use the test to see if the characteristics of the two populations are different. The significance tables describe the situation where the following assumptions are met.

(i) The two samples were drawn from two populations with the same distribution characteristics.
(ii) The two samples were drawn independently of each other.
(iii) The scores in each sample were drawn at random.
(iv) The scores are at least rankable.

A significant result indicates that at least one of these assumptions is not true. The test should not be used if assumptions (ii) and (iv) are not met and proper experimental procedure should guarantee assumption (iii). This leaves the possibility that the two populations from which the samples were drawn have different distribution characteristics. These could be differences in the mean, variance or skew of the populations or a combination of these. However, a significant result definitely indicates a difference in the medians of the populations sampled. More exactly, we may infer that if 2 scores are chosen at random (one from each group), the score from one group is significantly more likely to be larger (or smaller) than the score from the other group.

▶ *Example.*   University students about to begin a one year introductory course in statistics were assessed with an 'abstract reasoning' test. Two groups of students were selected for further study. One group was a random sample of 8 high scorers, while the other group was a random sample of 10 low scorers. Their results in an examination at the end of the course are given in the table below. The examination results were given on a 20 point scale (20 points is full marks). Do the results confirm our prediction that students who do well on the abstract reasoning task will also do well in the end of term examination?

DATA SUMMARY TABLE:

High scorers	4, 6, 8, 11, 11, 14, 17, 17	$n_1 = 8$
Low scorers	2, 2, 4, 5, 5, 6, 10, 11, 12, 12	$n_2 = 10$

1. Convert to ranks and find R, the sum of the ranks in the smaller group (group I)

High	3·5, 7·5, 9, 12, 12, 16, 17, 18	
Low	1·5, 1·5, 3·5, 5·5, 5·5, 7·5, 10, 12, 14·5, 14·5	$R = 95$

2. Calculate R' using formula (5.5)

$$R' = n_1(n_1 + n_2 + 1) - R = 8 \times 19 - 95 = 57$$

R' is the smaller.

Consulting table 5.5 we find for $n_1 = 8, n_2 = 10$ that $R' = 57$ is almost, but not quite, significant at the 5 per cent level of confidence using a 1-TAIL test.

Conclusion.   Our almost significant result could possibly be taken as evidence that high scorers on the abstract reasoning test produce higher scores in the statistics examination.

Table 5.5 Critical values of R or (R) in the Mann–Whitney test.

N_1	N_2	2-TAIL 10% 1-TAIL 5%	5% 2.5%	2% 1%	0.2% 0.1%
3	3	$R \leq 6$			
	4	6			
	5	7	6		
	6	8	7		
	7	8	7	6	
	8	9	8	6	
	9	10	8	7	
	10	10	9	7	
4	4	11	10		
	5	12	11	10	
	6	13	12	11	
	7	14	13	11	
	8	15	14	12	
	9	16	14	13	
	10	17	15	13	10
	11	18	16	14	10
5	5	19	17	16	
	6	20	18	17	
	7	21	20	18	
	8	23	21	19	15
	9	24	22	20	16
	10	26	23	21	16
	11	26	24	22	17
	12	28	26	23	17
6	6	28	26	24	
	7	29	27	25	21
	8	31	29	27	22
	9	33	31	28	23
	10	35	32	29	24
	11	37	34	30	25
	12	38	35	32	25
	13	40	37	33	26
7	7	39	36	34	29
	8	41	38	35	30
	9	43	40	37	31
	10	45	42	39	33
	11	47	44	40	34
	12	49	46	42	35
	13	52	48	44	36
	14	54	50	45	37
8	8	51	49	45	40
	9	54	51	47	41
	10	56	53	49	42
	11	59	55	51	44
	12	62	58	53	45
	13	64	60	56	47
	14	67	62	58	48
	15	69	65	60	50
9	9	66	62	59	52
	10	69	65	61	53
	11	72	68	63	55
	12	75	71	66	57
	13	78	73	68	59
	14	81	76	71	60
	15	84	79	73	62
	16	87	82	76	64

Table 5.5—continued.

N_1	N_2	2-TAIL 10% 1-TAIL 5%	5% 2·5%	2% 1%	0·2% 0·1%
10	10	$R \leq 82$	78	74	65
	11	86	81	77	67
	12	89	84	79	69
	13	92	88	82	72
	14	96	91	85	74
	15	99	94	88	76
	16	103	97	91	80
	17	106	100	93	80
11	11	100	96	91	81
	12	104	99	94	83
	13	108	103	97	86
	14	112	106	100	88
	15	116	110	103	90
	16	120	113	107	93
	17	123	117	110	95
	18	127	121	113	98
12	12	120	115	109	98
	13	125	119	113	101
	14	129	123	116	103
	15	133	127	120	106
	16	138	131	124	109
	17	142	135	127	112
	18	146	139	131	115
	19	150	143	134	118
13	13	142	136	130	117
	14	147	141	134	120
	15	152	145	138	123
	16	156	150	142	126
	17	161	154	146	129
	18	166	158	159	133
	19	171	163	154	136
	20	175	167	158	139
14	14	166	160	152	137
	15	171	164	156	141
	16	176	169	161	144
	17	182	174	165	148
	18	187	179	170	151
	19	192	183	174	155
	20	197	188	178	159
15	15	192	184	176	160
	16	197	190	181	163
	17	203	195	186	167
	18	208	200	190	171
	19	214	205	195	175
	20	220	210	200	179
16	16	219	211	202	184
	17	225	217	207	188
	18	231	222	212	192
	19	237	228	218	196
	20	243	234	223	201
17	17	249	240	230	210
	18	255	246	235	214
	19	262	252	241	219
	20	268	258	246	223
18	18	280	270	259	237
	19	287	277	265	242
	20	294	283	271	247
19	19	313	303	291	267
	20	320	309	297	272
20	20	348	337	324	298

5.6 Unrelated *t* test

SCORES: ALL INDEPENDENT
VARIABLES: (a) NOMINAL; 2 CATEGORIES (GROUPS I, II)
             (b) NORMALLY DISTRIBUTED SCORES
DATA SUMMARY TABLE:

<table>
<tr><td colspan="2" align="center">Groups</td><td></td></tr>
<tr><td align="center">I</td><td align="center">II</td><td></td></tr>
<tr><td>$X_{1,1}$</td><td>$X_{1,2}$</td><td>Sample sizes</td></tr>
<tr><td>$X_{2,1}$</td><td>$X_{2,2}$</td><td>n_1, n_2</td></tr>
<tr><td>$X_{n1,1}$</td><td>$X_{n2,1}$</td><td></td></tr>
</table>

Notes
X_{ij} is the *i*th score in the *j*th group.
n_1 is the number of scores in the first group.
n_2 is the number of scores in the second group.

Procedure
1. Compute ΣX_1 and ΣX_1^2, the sum and sum of squares in the first group.
2. Compute ΣX_2 and ΣX_2^2, the sum and sum of squares in the second group.
3. Compute *t*

$$t = \frac{\Sigma X_1/n_1 - \Sigma X_2/n_2}{\sqrt{\left(\dfrac{\Sigma X_1^2 - [(\Sigma X_1)^2/n_1] + \Sigma X_2^2 - [(\Sigma X_2)^2/n_2]}{n_1 + n_2 - 2}\right)\left(\dfrac{n_1 + n_2}{n_1 n_2}\right)}} \tag{5.6}$$

4. *t* has $n_1 + n_2 - 2$ degrees of freedom.
5. The significance of *t* may be assessed by reference to table 2.7.5. Some critical values have been extracted and are given below. A 1-TAIL test is appropriate if the direction of the difference between the two group means was correctly predicted on a properly reasoned basis.

	2-TAIL	10%	5%	2%	0.2%
	1-TAIL	5%	2.5%	1%	0.1%
$df = 10$	$t \geq 1.81$		2.23	2.76	4.14
15	1.75		2.13	2.60	3.73
20	1.72		2.09	2.53	3.55
30	1.70		2.04	2.46	3.39
60	1.67		2.00	2.39	3.23
120	1.66		1.98	2.36	3.16

Inferences.   Our two groups are samples from two populations. We wish to know whether these populations have different means. *t* is a measure of the difference in the sample means. Table 2.7.5 describes the situation where the following assumptions are met:

 (i) The populations have the same means.
 (ii) The populations have the same variances.
(iii) The populations are both normally distributed.
(iv) The samples are drawn at random.

A significantly large value of *t* indicates that one of these assumptions is untrue. The test should not be used when assumptions (ii) and (iii) are not true and proper experimental procedure should guarantee assumption (iv). This leaves the implication that the population means are different.

▶ Example. This example compares the examination results of students of two levels of reasoning ability. Six students of high ability and 7 of low ability were selected at random. Their examination marks (on a scale 1–100) are given below. Do the results support our hypothesis that students of high reasoning ability will perform better in the end of course examination?

SCORES: ALL INDEPENDENT
VARIABLES: (a) REASONING ABILITY; NOMINAL, 2 CATEGORIES
             (b) EXAMINATION RESULTS; POSSIBLY NORMALLY DISTRIBUTED SCORES
DATA SUMMARY TABLE:

Ability groups	
High	Low
78	58
56	63
64	54
71	32
62	43
59	64
	62

$n_1 = 6,$ $n_2 = 7$

1. $\Sigma X_1 = 390,$ $\Sigma X_2^2 = 25\,682$

2. $\Sigma X_2 = 376,$ $\Sigma X_2^2 = 21\,062$

3. Using formula (5.6)

$$t = \frac{390/6 - 376/7}{\sqrt{\left(\dfrac{25\,682 - [(390)^2/6] + 21\,062 - (376^2/7)}{6 + 7 - 2}\right)\left(\dfrac{6 + 7}{6 \times 7}\right)}}$$

$$t = \frac{11 \cdot 29}{5 \cdot 8} = 1 \cdot 94$$

$$df = n_1 + n_2 - 2 = 6 + 7 - 2 = 11$$

Our value of t is significant at the 5 per cent level ($t = 1 \cdot 81$ for $df = 10$, 1-TAIL test).

Conclusion.   Our results are indicative of a difference between the examination performances of high and low reasoning ability students.

5.7 Chi square ($m \times k$) test

SCORES: ALL INDEPENDENT
VARIABLES: (a) NOMINAL; 3 OR MORE CATEGORIES ($A_1 \ A_2 \ldots A_k$)
(b) NOMINAL; 3 OR MORE CATEGORIES ($B_1 \ B_2 \ldots B_m$)
DATA SUMMARY TABLE:

		(a) Category				
		A_1	A_2	$\ldots$	A_k	
(b) Category	B_1	$O_{1,1}$	$O_{1,2}$	$\ldots$	$O_{1,k}$	R_1
	B_2	$O_{2,1}$	$O_{2,2}$	$\ldots$	$O_{2,k}$	R_2
	$\vdots$	$\vdots$	$\vdots$	$\ldots$	$\vdots$	$\vdots$
	B_m	$O_{m,1}$	$O_{m,2}$	$\ldots$	$O_{m,k}$	R_m
		C_1	C_2	$\ldots$	C_k	N

Notes
O_{ij} are the observed frequencies in the ith and jth column.
R_i are the row *totals*.
C_i are the column *totals*.
N is the grand total.
m is the number of rows.
k is the number of columns.

Procedure. The procedure is broadly similar to that for the $2 \times k$ design, which should be studied.
1. Compute the expected frequency for each cell by multiplying its row and column totals then dividing by the grand total.
2. Check that the table of expected frequencies has the *same* row and column totals as the data summary table.
3. Compute χ^2

$$\chi^2 = \sum \frac{O^2}{E} - N \tag{5.7}$$

4. Calculate the degrees of freedom associated with this value of χ^2

$$df = (m - 1)(k - 1)$$

5. Our value of χ^2 may now be assessed for significance by consulting table 2.7.4. The table below reproduces as much of the table as may commonly be needed. Our values of χ^2 must exceed that given in the table to be significant. 1-TAIL and 2-TAIL test considerations do not apply here.

Significance level	5%	2.5%	1.0%	0.1%
$df = 4$	$\chi^2 \geq 9.5$	11·1	13·3	18·5
6	12·6	14·5	16·8	22·5
8	15·5	17·5	20·1	26·1
9	16·9	19·0	21·7	27·9
12	21·0	23·3	26·2	32·9
16	26·3	28·8	32·0	39·3

Inferences. We may look upon the data as m samples of scores from populations where the only possible scores are k nominal categories. The column totals $(C_1, C_2, \ldots, C_k)$ are estimates of the relative frequencies of scores in these k nominal categories. χ^2 measures how closely the relative frequencies in the m samples resemble the column totals.

Table 2.7.4 specifies how frequently certain values of χ^2 are exceeded when the following assumptions are met:

(i) The populations, from which our m samples are drawn, contain the *same* relative frequencies of scores in the k categories $(A_1, A_2, \ldots, A_k)$.
(ii) The m samples were drawn independently.
(iii) The scores within each sample were drawn at random.
(iv) The samples are of adequate size; see below.

Significantly large values of χ^2 indicate that at least one of these assumptions is not true. The test should not be used when assumptions (ii) and (iv) are not met by the data and proper experimental procedure should guarantee assumption (iii). This leaves the implication that our m samples were *not* in fact drawn from populations with the same relative frequencies of scores in the k categories $(A_1, A_2, \ldots, A_k)$.

Sample sizes. In this test, the χ^2 values given in significance tables are only approximately valid. For large sample sizes, the approximation is excellent. It is advisable *not* to use this test if more than 20 per cent of the expected frequencies are less than 5.

▶ Example. Students from four disciplines (Biology, Economics, Psychology and Sociology) attend an introductory statistics course. At the end of the course (which is not examined) they are allowed three options, (a) to proceed to a more advanced course, (b) to repeat the same course or (c) to attend no more statistics courses. The frequency with which each option is taken by students in a random sample from each discipline is given in the table below. Is there any evidence to show that the options taken by students depend upon their discipline?

SCORES;   ALL INDEPENDENT
VARIABLES: (a) DISCIPLINE; NOMINAL, 4 CATEGORIES $m = 4$
            (b) OPTION; NOMINAL, 3 CATEGORIES $k = 3$
DATA SUMMARY TABLE:

Discipline	Option Advanced course	Repeat course	No more courses	Totals
Biology	6	4	24	34
Economics	14	9	18	41
Psychology	8	7	26	41
Sociology	2	12	36	50
	30	32	104	166

1. Compute the expected frequencies

			Row totals
$\dfrac{34 \times 30}{166} = 6\cdot14$	$\dfrac{34 \times 32}{166} = 6\cdot55$	$\dfrac{34 \times 104}{166} = 21\cdot30$	34
$\dfrac{41 \times 30}{166} = 7\cdot41$	$\dfrac{41 \times 32}{166} = 7\cdot90$	$\dfrac{41 \times 104}{166} = 25\cdot69$	41
$\dfrac{41 \times 30}{166} = 7\cdot41$	$\dfrac{41 \times 32}{166} = 7\cdot90$	$\dfrac{41 \times 104}{166} = 25\cdot69$	41
$\dfrac{50 \times 30}{166} = 9\cdot04$	$\dfrac{50 \times 32}{166} = 9\cdot64$	$\dfrac{50 \times 104}{166} = 31\cdot33$	50
Column totals 30	32	104	166

2. Compute for each cell

$$\frac{O^2}{E}$$

5·86	2·44	27·04	
26·45	10·25	12·61	
8·63	6·20	26·31	
0·44	14·93	41·37	$\sum \dfrac{O^2}{E} = 202\cdot5$

3. Compute χ^2

$$\chi^2 = \sum \frac{O^2}{E} - N = 202\cdot5 - 166 = 16\cdot5$$

4. Calculate degrees of freedom

$$df = (m - 1)(k - 1) = (4 - 1)(3 - 1) = 3 \times 2 = 6$$

5. Our value of χ^2 is better than the critical value at the 2·5 per cent level (14·45) for 6 degrees of freedom.

Conclusion. The pattern of option preference across the 4 groups is significantly different.

5.8 Kruskal–Wallis test

SCORES: INDEPENDENT
VARIABLES: (a) NOMINAL; 3 OR MORE GROUPS $(G_1, G_2, \ldots, G_k)$
             (b) RANKABLE SCORES
DATA SUMMARY TABLE:

	Groups			
G_1	G_2	...	G_k	
x	x		x	n_1
x	x		x	n_2
x	x		x	$\vdots$
x	x		x	$\cdot$
	x		x	n_k
	x			$N = n_1 + n_2 + \ldots + n_k$

Notes
x are all independent rankable scores.
n_i is the number of scores in the ith group.
N is the total number of scores.
k is the number of groups.

Procedure
1. Rank all scores irrespective of group, i.e., from 1 to N.
2. Find the sum of the ranks for each group $R_1, R_2, \ldots, R_k$.
3. Calculate K

$$K = \sum \frac{R_j^2}{n_j} = \frac{R_1^2}{n_1} + \frac{R_2^2}{n_2} + \ldots + \frac{R_k^2}{n_k} \tag{5.8.1}$$

4. *Small samples.* When there are only 3 groups and less than 5 scores in each group, table 5.8 may be used to assess the significance of K. K must be larger than the figure given in the table.
5. Otherwise, we may compute χ^2

$$\chi^2 = \frac{12K}{N(N+1)} - 3(N+1) \tag{5.8.2}$$

with $(k-1)$ degrees of freedom.

The significance of χ^2 may be assessed with reference to table 2.7.4. Those values most likely to be required are given below. 1- and 2-TAIL considerations do not apply here.

Significance level		5%	2·5%	1%	0·1%
Number of groups	Degrees of freedom				
3	2	$\chi^2 \geq 6.0$	7·4	9·2	13·8
4	3	7·8	9·4	11·3	16·3
5	4	9·5	11·1	13·3	18·5
6	5	11·1	12·8	15·1	20·5
7	6	12·6	14·5	16·8	22·5
8	7	14·1	16·0	18·5	24·3

Inferences. Our k groups may be viewed as random samples from k different populations. The aim of the test is to see whether these populations have different distribution characteristics, in particular whether they have different means. The value K is a measure of the variability of the group rank totals. When the populations have different means, we expect K to be large.

The significance tables describe the situation where the following assumptions are met:

(i) The populations from which the samples were drawn have similar distribution characteristics.
(ii) The samples were drawn independently of each other.
(iii) The scores in the samples were drawn at random.

A significantly large value of K indicates that at least one of these assumptions is not true. If assumption (ii) is not met, the test should not be used and good experimental procedure should guarantee assumption (iii). This leaves the possibility that the distribution characteristics of the sampled populations are different. With 3 or more groups, a significant value of K almost certainly implies that the medians and possibly the means of the distributions are different. It is not possible, without applying further tests, to be more specific than this.

▶ Example. Students from four departments (Physics, Biology, Psychology and Sociology) attended an introductory course in statistics. Random samples of students from the departments were given a short test without notice. The results were given in terms of a rank order for each student. These are given below. Is there evidence that students from different departments have a different aptitude for statistics?

SCORES: ALL INDEPENDENT
VARIABLES: (a) DEPARTMENT; NOMINAL, 4 CATEGORIES $k = 4$
             (b) EXAMINATION RESULT; RANK ORDER SCORES
DATA SUMMARY TABLE:

Physics	Biology	Psychology	Sociology	
1	5	2	6	$n_1 = 3$
4	9	3	11	$n_2 = 3$
10	12	7	13	$n_3 = 4$
		8	14	$n_4 = 4$
				$N = 14$

1. We may immediately compute the sum of ranks for each group

$$R_1 = 15, \qquad R_2 = 26, \qquad R_3 = 20, \qquad R_4 = 44$$

2. Calculate K using formula (5.8.1)

$$K = \frac{15^2}{3} + \frac{26^2}{3} + \frac{20^2}{4} + \frac{44^2}{4} = 884\cdot33$$

3. Calculate χ^2 using formula (5.8.2)

$$\chi^2 = \frac{12 \times 884\cdot33}{14 \times 15} - 3 \times 15 = 5\cdot533; \qquad df = 4 - 1 = 3$$

4. We find that our value of χ^2 is not significant.

$$(df = 3, \qquad \chi^2 = 7\cdot8 \text{ required at 5 per cent level})$$

Conclusion. The results do not permit us to conclude that students in the four departments differ in their aptitude for statistics.

Table 5.8   Critical values of K in the Kruskal–Wallis test.

Significance level			5%	2·5%	1%	0·1%
n_1	n_2	n_3				
3	2	2	$K \geq$ 134·00			
3	3	1	136·00			
3	3	2	194·17	195·33		
3	3	3	267·00	269·67	279·00	
4	2	1				
4	2	2	194·00	195·00		
4	3	1	193·25	197·00		
4	3	2	265·83	270·00	273·33	
4	3	3	355·58	358·92	364·33	
4	4	1	262·25	271·25	275·00	
4	4	2	352·50	360·50	367·00	
4	4	3	457·58	466·33	474·58	494·00
4	4	4	581·00	594·00	606·50	627·00
5	2	1	192·00			
5	2	2	263·70	270·00	274·00	
5	3	1	262·20	270·33		
5	3	2	350·63	357·53	365·83	
5	3	3	458·13	465·47	473·87	492·00
5	4	1	348·20	356·20	366·25	
5	4	2	454·00	462·75	475·25	490·50
5	4	3	580·53	590·33	603·78	621·33
5	4	4	722·80	738·20	754·70	776·05

5.9   Kendall's test for ranked categories

SCORES: ALL INDEPENDENT
VARIABLES: (a) RANKABLE CATEGORIES $(Q_1 < Q_2 < \ldots < Q_k)$
                     (b) RANKABLE CATEGORIES $(R_1 < R_2 < \ldots < R_m)$
DATA SUMMARY TABLE:

(a) Rankable categories

$$Q_1 < Q_2 < \ldots < Q_k$$

	$Q_1 < Q_2 < \ldots < Q_k$			
R_1	F	F	$\ldots$	F
R_2	F	F	$\ldots$	F
$\vdots$	$\vdots$	$\vdots$	$\ldots$	$\vdots$
R_m	F	F	$\ldots$	F

(b) Rankable categories

$N = \Sigma F$

Notes
F are *frequencies* of scores which fall in the relevant row and column categories.
N is the total number of scores, the sum of the frequencies.
m is the number of rows.
k is the number of columns.

Procedure
1. For *each* frequency compute a value s (there will be $k.m$ such values): $s = F \times (B.L. - B.R.)$, where
   $B.R.$ is the total number of scores (sum of frequencies) *below* and to the *right* of that F in the table, and
   $B.L.$ is the total number of scores *below* and to the *left* of that F in the table.
2. Compute S by summing all individual values of s based on every value of F in the table. $S = \Sigma s$.
3. We may assess the significance of S by calculating

$$Z = \frac{S}{\sqrt{\left(\dfrac{(2N + 5)(N^2 - N)}{18}\right)}} \quad \text{or} \quad \frac{3S}{\sqrt{(\frac{1}{2}(2N + 5)(N^2 - N))}} \tag{5.9.1}$$

   which can be referred to tables of the normal distribution (table 2.7.2).
4. A value of Kendall's τ may be also computed if so desired.

$$\tau = \frac{S}{\frac{1}{2}N(N - 1)} \tag{5.9.2}$$

Values of τ lie along a scale from $-1\cdot0$ to $+1\cdot0$ and are roughly analogous to the Pearson product moment correlation coefficient.

Inferences.   In this test, we are concerned with the possibility that high ranking categories on scale (a) will be associated with high ranking categories on scale (b). A negative correlation ($S = -$ve large) will arise when high ranking categories on scale (a) are associated with *low* ranking categories on scale (b). When there is no or little correlation between the two scales, the value of S (or τ) will be near to zero. The significance test is based on the following assumptions:

 (i) There is no association between the two variables.
 (ii) All pairs of scores are drawn at random.
(iii) The categories are rankable.

A significantly large value of S indicates that one of these assumptions is not true. The test should not be used if assumption (iii) is not met and proper experimental procedure should guarantee assumption (ii). This leaves the possibility that the categories are in fact associated. This test becomes increasingly insensitive as the number of rows and columns decreases.

▶ **Example.** Fifty students were given an abstract reasoning test before beginning an introductory statistics course. The test involved solving four increasingly difficult problems in a short space of time. Each student was scored according to the number of correct solutions offered. The examination results at the end of the course were given as either 'fail', 'satisfactory' or 'good'. The frequency table below summarizes the results on both tests. Is there evidence for a positive correlation?

SCORES: INDEPENDENT
VARIABLES: (a) REASONING TEST: 5 RANKED CATEGORIES
                (0, 1, 2, 3, 4 problems solved)
           (b) EXAMINATION RESULT: 3 RANKED CATEGORIES
                (FAIL, SATISFACTORY, GOOD)
DATA SUMMARY TABLE:

Reasoning test (Questions answered)

Examination result		0	1	2	3	4	
	Good	0	3	4	3	2	
	Satisfactory	1	8	17	4	0	
	Fail	2	5	1	0	0	$N = 50$

1. For each frequency calculate

$$S = F(B.L. - B.R.)$$

Note (a) that scores in the left hand column have no scores below and left ($B.L.$);
       (b) that scores in the right hand column have no scores below and right ($B.R.$);
       (c) that scores in the bottom row have no scores below them, so $s = 0$

Top row s values

$$3(3 - 22), \quad 4(16 - 4), \quad 3(34 - 0), \quad 2(38 - 0)$$

Middle row s values

$$1(0 - 6), \quad 8(2 - 1), \quad 17(7 - 0), \quad 4(8 - 0)$$

2. Calculate $S = \sum s$

$$S = (3 \times -19) + (4 \times 12) + (3 \times 34) + (2 \times 38)$$
$$+ (1 \times -6) + (8 \times 1) + (17 \times 7) + (4 \times 8)$$
$$= 322$$

3. Calculate Z

$$Z = \frac{3S}{\sqrt{(\frac{1}{2}(2N + 5)(N^2 - N))}} = \frac{3 \times 322}{\sqrt{(\frac{1}{2}(2 \times 50 + 5)(50^2 - 50))}} = \frac{966}{358}$$
$$= 2 \cdot 698$$

Since a positive value of S was predicted, a 1-TAIL test is appropriate. We find that our Z value of $2 \cdot 69$ is significant at better than the 1 per cent level ($Z = 2 \cdot 33$ at 1 per cent level).

Conclusion. Our results provide evidence that the results of the reasoning test are correlated with the statistics examination results.

5.10   Kendall's test for rankable scores

SCORES: ALL INDEPENDENT
VARIABLES: (a) RANKABLE SCORES $(a_1, a_2, a_3, \ldots, a_n)$
            (b) RANKABLE SCORES $(b_1, b_2, b_3, \ldots, b_n)$
DATA SUMMARY TABLE:

Variable (a)	a_1	a_2	a_3	...	a_N
Variable (b)	b_1	b_2	b_3	...	b_N

Notes

a_i, b_i are pairs of rankable scores.

The *a* scores have been put in rank order for the purpose of the table so that $a_1 < a_2 < \ldots < a_N$. The (*a*, *b*) pairs have been kept together.

N is the number of pairs of scores.

Procedure

1. Check that the scores on the top row of the table are in rank order.
2. Take the first score on the bottom row b_1 and compare it with all other scores to the right of it, $b_2, b_3, \ldots, b_n$.
   Whenever a score to the right is larger than it, score $+1$ point.
   Whenever a score to the right is smaller than it, score -1 point.
   If the scores are the same, score 0 points.
3. Repeat this process for each score on the bottom row. Remember to compare it only with scores to the right.
   Note: If two *a* scores (on the top row) have the same rank, then do not compare their *b* scores (i.e., 0 points for that comparison).
4. The total number of points is called *S*. The significance of *S* may be found by referring it to table 5.10. *S* (either positive or negative) must be equal to or bigger than the value in the table to be significant. A 1-TAIL test is appropriate if the direction of the correlation (positive or negative *S*) was predicted before the results were available.
5. For large samples, we may compute

$$Z = \frac{S}{\sqrt{\left(\dfrac{(2N + 5)(N^2 - N)}{18}\right)}} \quad \text{or} \quad \frac{3S}{\sqrt{(\frac{1}{2}(2N + 5)(N^2 - N))}} \tag{5.10.1}$$

which can be referred to tables of the normal distribution (table 2.7.2). Some critical values are given below.

	10%	5·0%	2%	0·2%
2-TAIL				
1-TAIL	5%	2·5%	1%	0·1%
$Z \geq$	1·64	1·96	2·33	3·10

6. While *S* is adequate for a significance test, we may also compute τ as a measure of association analogous to Pearson's product moment correlation.

$$\tau = \frac{S}{\sqrt{(\frac{1}{2}N(N - 1))}} \tag{5.10.2}$$

Table 5.10   Critical values of S for Kendall's test of rank correlation.

N	2-TAIL 10% / 1-TAIL 5%	5% / 2.5%	2% / 1%	0.2% / 0.1%
4	$S \geq 6$			
5	8	10	10	
6	11	13	13	
7	13	15	17	21
8	16	18	20	24
9	18	20	24	30
10	21	23	27	35
11	23	27	31	39
12	26	30	36	44
13	28	34	40	50
14	33	37	43	55
15	35	41	49	61
16	38	46	52	68
17	42	50	58	74
18	45	53	63	81
19	49	57	67	87
20	52	62	72	94
21	56	66	78	102
22	61	71	83	109
23	65	75	89	115
24	68	80	94	124
25	72	86	100	132
26	77	91	107	139
27	81	95	113	147
28	86	100	118	156
29	90	106	126	164
30	95	111	131	171
31	99	117	137	181
32	104	122	144	188
33	108	128	152	198
34	113	133	157	207
35	117	139	165	215
36	122	146	172	226
37	128	152	178	234
38	133	157	185	243
39	139	163	193	253
40	144	170	200	264
41	148	176	208	274
42	153	183	215	283
43	159	189	223	293
44	164	196	230	304
45	170	202	238	314

Inferences. S measures the extent to which the rank ordering of the (b) variable matches the rank ordering of the (a) variable. Large positive values of S indicate a positive correlation. Large negative values of S indicate a negative correlation. Table 5.10 describes the situation where the following assumptions are met:

(i) There is no correlation between the two variables.
(ii) Each pair of scores is independent of the other pairs.
(iii) Each pair of scores is drawn at random.

A significant value of S indicates that one of these assumptions is not true. This test should not be used if assumption (ii) is not met and proper experimental procedure should guarantee assumption (iii). This leaves the possibility that the two variables are correlated.

The correlation is one of rank order. Such a correlation implies that an increase in score on one variable will often be accompanied by an increase in score on the other variable. No inference can be made about the *size* of the increase. Moreover, the function relating the two variables is not necessarily linear.

The effect of tied observations is to make the value of S (and τ) slightly smaller and less significant. It is possible to correct for this effect but the improvement in accuracy is usually too small to warrant the effort.

▶ **Example.** Nine students were assessed for abstract reasoning ability on a test which rated them on a scale from 0 (poor) to 10 (good). These results are compared below with their examination results which were given on a scale from 0 (poor) to 7 (good). We expect a positive correlation between the two sets of scores. Is our expectation confirmed?

SCORES: ALL INDEPENDENT
VARIABLES: (a) REASONING ABILITY, RANKABLE SCORES (0–10)
            (b) EXAMINATION RESULTS, RANKABLE SCORES (0–7)
DATA SUMMARY TABLE:

Abstract reasoning	0	5	2	3	1	8	10	9	7
Examination result	2	4	2	3	0	2	6	7	5

1. The top row of the table is not in rank order. It must be rearranged

Reasoning	0	1	2	3	5	7	8	9	10
Examination	2	0	2	3	4	5	2	7	6

2. Comparisons

Score no. 1	X	-1	0	$+1$	$+1$	$+1$	0	$+1$	$+1$
no. 2		X	$+1$	$+1$	$+1$	$+1$	$+1$	$+1$	$+1$
no. 3			X	$+1$	$+1$	$+1$	0	$+1$	$+1$
no. 4				X	$+1$	$+1$	-1	$+1$	$+1$
no. 5					X	$+1$	-1	$+1$	$+1$
no. 6						X	-1	$+1$	$+1$
no. 7							X	-1	-1
no. 8								X	0

$$S = 20; \quad N = 9$$

3. Consulting table 5.10 we find that on a 1-TAIL test our value of S is significant at the 2·5 per cent level of significance.

4.
$$\tau = \frac{S}{\frac{1}{2}N(N-1)} = \frac{20}{\frac{1}{2} \times 9 \times 8} = \frac{20}{36} = 0\cdot56$$

Conclusion. A significant positive rank order correlation exists between scores on the reasoning task and examination performance.

6. Tests for two variables (Related scores)

6.1   General

The tests in this chapter mirror those in chapter 5. The special feature here is that the scores are not all independent. The simplest case of related scores occurs when each subject serves as his own control. In this case, two scores come from the same source and are related.

Tests which use related scores are usually more sensitive than other tests because they have the property of minimizing the effects of certain irrelevant factors by cancelling them out. This property has made some of these tests, especially the Wilcoxon and related t tests, very popular for small scale experimentation.

The Wilcoxon test is especially useful for simple experiments and has become enormously popular. Its main advantage is that there are apparently no requirements concerning the distribution of the scores. However, a note of caution must be issued since the test requires that the differences between the pairs of scores should be at least rankable. This may not seem to be a very severe restriction at first sight but many examples which I have seen do not meet this restriction. For example, a weak student improves his examination scores from 35 to 45 per cent after coaching. This is compared with a strong student who also improves his mark this time from 91 to 93 per cent. It is very debatable whether the improvement of the weaker student (10 per cent) is *better* than the improvement of the stronger student (2 per cent). We may note that it is impossible for the stronger student to improve his mark by 10 per cent. The alert investigator will notice many cases where difference scores cannot meaningfully be ranked across matched pairs.

The alternative is to use a sign-test which does not make this difficult assumption. Unfortunately, the sign test is considerably less sensitive than the Wilcoxon test. To fill this gap, a new test is offered (section 6.6). It requires a little more information than is present in the data for a Wilcoxon test since each score is replaced by a set of scores (hence the title 'repeated measures'). When this information is available we have a 2 group test which makes comparisons only within sets of related scores. In this respect, it is very similar to the Friedman test.

6.2 McNemar's test

SCORES: RELATED; MATCHED PAIRS ACROSS VARIABLE a
VARIABLES: (a) NOMINAL; 2 CATEGORIES (GROUPS I, II)
                 (b) NOMINAL; 2 CATEGORIES (X, Y)
DATA SUMMARY TABLE:

Matched pair	1	2	3	4	5	6	7	8	...	n
Group I	x	x	x	y	y	x	x	x	...	x
Group II	x	y	y	y	y	y	y	x	...	y

Note
n is the number of pairs of related scores.

Procedure
1. Count the number of (x, y) pairs with x in group I and y in group II; A.
2. Count the number of (y, x) pairs with y in group I and x in group II; B.
3. Calculate $N = A + B$. N is the number of pairs showing changes.
4. The significance of A (or B, whichever is the smaller) can be assessed by consulting table 4.2. Your value must be smaller than that given in the table. If a successful prediction was made whether or not A would prove larger than B, a 1-TAIL test is appropriate.
5. For large samples, calculate Z

$$Z = \frac{A - B}{\sqrt{(A + B)}} \quad \text{or} \quad \frac{B - A}{\sqrt{(A + B)}} \tag{6.2}$$

which can be assessed by reference to table 2.7.2. Some critical values have been extracted and are given below.

	10%	5%	2%	0.2%
2-TAIL	10%	5%	2%	0.2%
1-TAIL	5%	2.5%	1%	0.1%
$Z \geq$	1.64	1.96	2.33	3.10

Inferences. This test focuses on those pairs of scores which show a difference across the two groups. $(A - B)$ is a measure of whether the changes were predominantly in one direction. Table 4.2 and the approximation to the normal distribution describe the situation where the following assumptions are met:

 (i) The pairs of scores were drawn from a population where (x, y) pairs are as common as (y, x) pairs.
(ii) The pairs were drawn at random.

A significantly large (or small) value of A indicates that one of these assumptions is not true. Proper experimental procedure should guarantee assumption (ii). This leaves the possibility that one kind of difference pair is more common in the population from which the sample of pairs was drawn. This in turn implies that groups I and II are themselves drawn from populations with different distribution characteristics. For example, this may imply that the population from which group I is drawn contains more x (or y) scores than the population represented by group II.

▶ Example.   After the examination at the end of an introductory course on statistics, we chose 19 pairs of students, each pair matched for examination performance. The pairs contained one student from the Biology department and one from the Education department. The students were all asked whether they felt that the course was relevant to their particular needs (Yes or No). Do the results indicate any difference between students in the two departments?

SCORES: RELATED; PAIRS MATCHED ACROSS DEPARTMENT
VARIABLES: (a) DEPARTMENT; NOMINAL, 2 CATEGORIES (BIOLOGY, EDUCATION)
                    (b) ASSESSED RELEVANCE; NOMINAL, 2 CATEGORIES (YES, NO)
DATA SUMMARY TABLE: Y = Yes; N = NO

Pair	1	2	3	4	5	6	7	8	9	10	11	12	13	14	15	16	17	18	19	
Biology	Y	N	Y	Y	N	Y	Y	N	N	Y	Y	Y	Y	Y	Y	N	N	Y	Y	
Education	N	N	N	Y	Y	N	N	N	N	N	N	N	N	N	N	Y	N	N	N	
Y, N	✓		✓			✓	✓			✓	✓	✓	✓	✓	✓			✓	✓	$A = 1$
N, Y				✓												✓				$B = 2$

Procedure
1. $A = 12$; number of Yes/No pairs.
2. $B = 2$; number of No/Yes pairs.
3. $N = A + B = 14$.
4. Consulting table 4.2 we find that our result ($B = 2$, $N = 14$) is significant at better than the 2·0 per cent level (2-TAIL test). A 2-TAIL test is used because no prediction was made concerning who would find the course most relevant to their needs.

Conclusion.   Biology students are more likely to find the course relevant than Education students.

6.3 Cochran's Q test

SCORES: RELATED; SETS MATCHED ACROSS VARIABLE a
VARIABLES: (a) NOMINAL; MORE THAN 2 CATEGORIES (GROUPS I, II, ..., k)
(b) NOMINAL; 2 CATEGORIES (1, 0)
DATA SUMMARY TABLE:

	Group				
Set	I	II	...	k	
1	X	X	...	X	R_1
2	X	X	...	X	R_2
3	X	X	...	X	R_3
$\vdots$	$\vdots$	$\vdots$		$\vdots$	
n	X	X		X	R_n
	C_1	C_2	...	C_k	T

Notes

X are scores on the (b) variable and may take only values 0 or 1 (e.g., No = 0; Yes = 1).
k is the number of groups.
n is the number of sets of related scores.
R are the row totals (R is always equal to or less than k).
C are the column totals (C is always equal to or less than n).
T is the grand total $= \Sigma R$ or ΣC.

Procedure

1. Find ΣC^2 and ΣR^2.
2. Compute Q

$$Q = \frac{(k-1)(k\Sigma C^2 - T^2)}{kT - \Sigma R^2} \tag{6.3}$$

3. Q closely approximates chi square with $(k-1)$ degrees of freedom and may be referred to table 2.7.4.
4. Some critical values are given below. The approximation of Q to the chi square distribution is very good for large values of n (number of sets of scores). For small values, however, the use of critical chi square values may result in a very conservative (i.e., lower power) test. No exact tables exist for small values of n.

Significance level		5%	2.5%	1%	0.1%
Number of groups					
$k = 3$	$df = 2$	$\chi^2 \geq$ 6.0	7.4	9.2	13.8
4	3	7.8	9.3	11.3	16.3
5	4	9.5	11.1	13.3	18.5
6	5	11.1	12.8	15.1	20.5

Inferences. We may regard our results as k samples from populations where the only possible scores are 0 and 1. We wish to know whether the relative frequencies of these two types of scores are the same in the different populations. Q is a measure of the variability of these relative frequencies across samples.

The chi square approximation describes the situation where the following assumptions are met:

(i) The relative frequency of scores 0, 1 is the same in the k populations sampled.
(ii) The n sets of scores were chosen randomly.

A significantly large value of Q indicates that one of these assumptions is untrue. Good experimental procedure should guarantee assumption (ii). This leaves the implication that the relative frequency of scores 0, 1 is different from population to population. It is not possible to produce a more specific inference than this.

▶ Example. It was decided to check whether students in different departments were equally happy that the course in introductory statistics was likely to satisfy their future needs. The students had recently sat the end of course examination and it was feared that the students' response might be influenced by the mark they had received. The students were accordingly divided into sets with one student from each department per set. Students in each set were matched so that they had roughly equivalent examination marks. Each student was asked whether or not he believed that the course would satisfy his future needs. Do the results provide evidence that students from different departments have different views?

SCORES: RELATED: SETS MATCHED ACROSS DEPARTMENTS
VARIABLES: (a) DEPARTMENT; NOMINAL, 4 CATEGORIES
  (b) OPINION; NOMINAL, 2 CATEGORIES (Yes = 1, No = 0)
DATA SUMMARY TABLE:

Set	Biology	Economics	Psychology	Education	R
1	1	0	1	0	2
2	0	0	0	0	0
3	1	1	1	0	3
4	1	0	1	1	3
5	0	0	1	0	1
6	1	0	0	0	1
7	1	1	1	1	4
8	1	0	0	0	1
9	0	1	0	0	1
10	1	0	1	0	2
11	0	0	1	0	1
C	7	3	7	2	T = 19

$k = 4$

Procedure
1. $\Sigma C^2 = 7^2 + 3^2 + 7^2 + 2^2 = 111.$

   $\Sigma R^2 = 2^2 + 3^2 + 3^2 + 1^2 + 1^2 + 4^2 + 1^2 + 1^2 + 2^2 + 1^2 = 47.$

2. $$Q = \frac{(k-1)(k\Sigma C^2 - T^2)}{kT - \Sigma R^2} = \frac{2(3 \times 111 - 19^2)}{3 \times 19 - 47} = 5{\cdot}6$$

   $\chi^2 = Q = 5{\cdot}6 \qquad df = k - 1 = 4 - 1 = 3$

3. We find that our value of χ^2 is not significant even at the 5 per cent level ($\chi^2 = 6{\cdot}0$ for $df = 3$ at 5 per cent level).

Conclusion. The evidence is not strong enough to support the idea that there are differences across departments in the extent to which students feel that the course will satisfy their future needs.

6.4   Sign test

SCORES: RELATED; MATCHED PAIRS ACROSS VARIABLE a
VARIABLES: (a) NOMINAL; 2 CATEGORIES (GROUPS I, II)
             (b) RANKABLE CATEGORIES $(R_1 < R_2 < \ldots < R_3)$
DATA SUMMARY TABLE:

		Groups	
		I	II
Matched pairs	1	R	R
	2	R	R
	3	R	R
	.	.	.

Note
R are rankable categories or scores.

Procedure

1. Find A where:
   A is the number of pairs in which the *Group I* score is *higher* than the group II score.
2. Find B where:
   B is the number of pairs in which the *Group I* score is the lower of the two (ignore pairs which tie).
3. Find N: $N = A + B$.
4. To assess the significance of A (or B *whichever is the lower*) consult table 4.2. The value of A (or B) must be equal to or lower than the value in the table to be significant. If a correct prediction was made as to which (A or B) would be the smaller, then a 1-TAIL test is appropriate.
5. For large samples, we may compute Z:

$$Z = \frac{A - B}{\sqrt{(A + B)}} \quad \text{or} \quad \frac{B - A}{\sqrt{(A + B)}} \tag{6.4}$$

which may be assessed by reference to table 2.7.2. Some critical values have been extracted and are given below.

| 2-TAIL | 10% | 5% | 2% | 0.2% |
1-TAIL	5%	2.5%	1%	0.1%
$Z \geq$	1.64	1.96	2.33	3.10

Inferences.   Groups I and II may be regarded as samples from two populations which may have different distribution characteristics (mean, variance, skew, etc.). If they have the same distribution characteristics, then a pair of scores drawn at random one from each population is expected to be equal. Scores from one distribution will be higher as often as lower. Table 4.2 describes the situation where the following assumptions are met:

(i) The two populations from which the samples are drawn have the same distribution characteristics.
(ii) The pairs of scores have been drawn at random.

A significantly small value of A (or B) indicates that one of these is not true. Good experimental procedure should guarantee assumption (ii). This leaves the implication that the distribution characteristics differ. This usually indicates that the medians and possibly the means of the two populations are different.

► Example.   Students from various departments about to begin an introductory course in statistics are all given an abstract reasoning test which grades them as 'high' or 'low' ability. We wish to relate these two grades with the student end of course examination result. It is possible, however, that students in different departments might work with different levels of motivation. To avoid this effect we select at random pairs of students where both members of the pair are from the same department but one has 'high' and the other 'low' reasoning ability. The end of term examination was marked on a 4 point scale, 1 = fail, 2 = poor, 3 = satisfactory, 4 = good. Do the results of this experiment support our prediction that 'high' reasoning ability will be associated with better examination results?

SCORES: RELATED; PAIRS MATCHED ACROSS ABILITY GROUPS
VARIABLES: (a) REASONING ABILITY; NOMINAL, 2 GROUPS (HIGH, LOW)
                    (b) EXAMINATION RESULTS; 4 RANKED CATEGORIES
DATA SUMMARY TABLE:

| Matched pairs | Reasoning ability Groups | | B | A |
	Low	High		
1	2	3	✓	
2	3	3		—
3	2	4	✓	
4	1	3	✓	
5	3	1		✓
6	2	2		—
7	1	4	✓	
8	2	3	✓	
9	2	4	✓	
10	1	2	✓	
11	2	3	✓	
12	2	4	✓	
13	3	4	✓	
14	3	4	✓	
15	2	3	✓	
			12	1

Procedure
1. $A = 1$.
2. $B = 12$.
3. $N = A + B = 13$.
4. Consulting table 4.2 we find that our value of A is significant at the 1 per cent level (1-TAIL test).
   We use a 1-TAIL test because it was predicted that the 'high' group would have better examination scores (i.e., that B would be larger than A).

Conclusion.   High abstract reasoning ability is associated with better scores in the statistics examination.

6.5 Wilcoxon's matched pairs test

SCORES: RELATED; PAIRS MATCHED ACROSS VARIABLE a
VARIABLES: (a) NOMINAL; 2 CATEGORIES (GROUPS I, II)
  (b) RANKABLE SCORES

DATA SUMMARY TABLE:

	Pairs	1	2	...	m
Groups	I	X	X	...	X
	II	X	X	...	X
	D	D_1	D_2		D_n
Rank of $|D|$		r_1	r_2		r_n

Notes

m is the number of pairs of scores.

D is the difference between each pair of scores.

n is the number of *non-zero* differences.

r is the rank of $|D|$.

Procedure

1. Ignore all zero values of D. Count (n) the number of *non-zero* values of D.
2. Rank the non zero values of D *irrespective of sign*: small values of D have small ranks (e.g., 1, 2, . . .).
3. Find R the sum of the ranks associated with negative D's
  R' the sum of the ranks associated with positive D's.
4. Find T the smaller value of R, R'.
5. The significance of T can be assessed by reference to table 6.5.
  R must be smaller than the value given in the table to be significant.
  A 1-TAIL test is appropriate following a correct prediction of the direction of the overall difference between groups I, II.
6. For *large samples*, we may compute Z:

$$Z = \frac{[n(n+1)/4] - T}{\sqrt{\left(\frac{n(n+1)(2n+1)}{24}\right)}} \qquad (6.5)$$

which is assessed by reference to table 2.7.2. Some critical values have been extracted and are given below.

2-TAIL	10%	5%	2%	0·2%
1-TAIL	5%	2·5%	1%	0·1%
$Z \geq$	1·64	1·96	2·33	3·10

Inferences. We may regard our two groups as samples drawn from two populations. We wish to know whether the populations have different means. Large differences between the means will lead to small values of R. Table 6.5 describes the situation where the following assumptions are met:

 (i) The two populations from which the samples were drawn have the same distribution characteristics (mean, variance, skew, etc.).
 (ii) The pairs of scores are drawn at random.
(iii) It is meaningful to rank the difference scores.

A significantly small value of R indicates that one of these assumptions is untrue. Proper experimental procedure should guarantee assumption (ii). The test should not be used if assumption (iii) is not valid. This leaves the implication that the distribution characteristics are different. The small value of R is most likely caused by differences in the means but differences in skew will also affect R.

▶ Example. Students from a number of college departments were given an abstract reasoning test before beginning an introductory course in statistics. This test grades them as 'high' or 'low' in reasoning ability. It was expected that students high on the scale would perform better in the end of course examinations. It is possible, however, that students in different departments might work with different levels of motivation. To

Table 6.5 Critical values of *T* in the Wilcoxon test.

	2-TAIL 10% 1-TAIL 5%	5·0% 2·5%	2% 1%	0·2% 0·1%
N = 5	*T* ≤ 0			
6	2	0		
7	3	2	0	
8	5	3	1	
9	8	5	3	
10	10	8	5	0
11	13	10	7	1
12	17	13	9	2
13	21	17	12	4
14	25	21	15	6
15	30	25	19	8
16	35	29	23	11
17	41	34	27	14
18	47	40	32	18
19	53	46	37	21
20	60	52	43	26
21	67	58	49	30
22	75	65	55	35
23	83	73	62	40
24	91	81	69	45
25	100	89	76	51
26	110	98	84	58
27	119	107	92	64
28	130	116	101	71
30	151	137	120	86
31	163	147	130	94
32	175	159	140	103
33	187	170	151	112

avoid this complication matched pairs of students were used. Each pair came from the same department but contained one high and one low in abstract reasoning ability. Twelve pairs of students were used. Their examination results, given on a 10 point scale (0 = lower 10 per cent, 9 = top 10 per cent) are shown below.

SCORES: RELATED; PAIRS MATCHED ACROSS ABILITY GROUPS
VARIABLES: (a) ABILITY GROUPS; NOMINAL, 2 CATEGORIES
                  (b) EXAMINATION RESULT; RANKABLE SCORE
DATA SUMMARY TABLE:

	Pairs	1,	2,	3,	4,	5,	6,	7,	8,	9,	10,	11,	12,
	High	6	9	5	7	3	5	4	9	4	8	8	6
Ability groups	Low	2	6	5	4	4	0	6	3	3	3	4	6
	D	4	3	0	3	−1	5	−2	6	1	5	4	0
	Rank of |*D*|	6·5	4·5	—	4·5	1·5	8·5	3	10	1·5	8·5	6·5	—

Procedure

1. There are 10 non-zero values of *D*: *N* = 10.
2/3. The sum of the ranks associated with *negative D*'s will clearly be smaller

$$R = 1·5 + 3 = 4·5 = T$$

4. Consulting table 6.5, we find that our value of *T* is significant at better than 1 per cent level (1-TAIL test). We use a 1-TAIL test because it was predicted that the high ability group would perform better.
5. If we use our large sample formula, we obtain

$$Z = \frac{[n(n+1)/4] - T}{\sqrt{\left(\frac{n(n+1)(2n+1)}{24}\right)}} = \frac{[10 \times 11/4] - 4·5}{\sqrt{\left(\frac{10 + 11 \times 21}{24}\right)}} = 2·34$$

We find that our *Z* value is significant at just better than 1 per cent level on a 1-TAIL test. This is the same result obtained in (4) above.

Conclusion. Our significantly low value of *T* suggests that high abstract reasoning ability is associated with good examination results.

6.6 Two groups repeated measures test

SCORES: RELATED; SETS MATCHED ACROSS VARIABLE a
VARIABLES: (a) NOMINAL; 2 CATEGORIES (GROUPS I, II)
           (b) RANKABLE SCORES
DATA SUMMARY TABLE:

		Groups	
Matched sets		I	II
1		$X_1 \ldots X_{n1}$	$X_1 \ldots X_{n2}$
2		$X_1 \ldots X_{n1}$	$X_1 \ldots X_{n2}$
$\vdots$		$\vdots$	$\vdots$
m		$X_1 \ldots X_{n1}$	$X_1 \ldots X_{n2}$

Notes

X_i are rankable scores.

n_1 is the number of scores *per set* in group 1.

n_2 is the number of scores *per set* in group 2.

n_1 is always equal to or less than n_2 for computational convenience.

m is the number of matched sets.

Procedure

1. For each set, rank the scores *within the set* from 1 to $(n_1 + n_2)$.
2. Find the sum of the ranks in group I: R_1.
3. Compute Z

$$Z = \frac{mn_1(n_1 + n_2 + 1) - 2R_1}{\sqrt{(\frac{1}{3}mn_1n_2(n_1 + n_2 + 1))}}$$ (6.6)

Note: When $n_1 = n_2 = 1$ use the sign test (6.4) where the exact tables will be more accurate.

Z may be assessed with reference to table 2.7.2. Some critical values have been extracted and are given below. A 1-TAIL test is appropriate when the direction of the overall difference between the groups was predicted in advance.

2-TAIL	10%	5%	2%	0·2%
1-TAIL	5%	2·5%	1%	0·1%
$Z \geq$	1·64	1·96	2·33	3·10

Inferences.   The scores in groups I and II can be considered as samples from two populations. We are interested to know whether the populations have different means. Formula (6.6) and table 2.7.2 deal with the situation where the following assumptions are met:

(i) The two populations from which our samples are drawn, have similar distribution characteristics.

(ii) The samples were drawn at random, within the limitations of the matching procedure.

A significantly small value of R indicates that one of these assumptions is not true. Good experimental procedure should guarantee assumption (ii). This leaves the implication that the population distribution characteristics are different. Different means or differences in skew might be involved. The effect is most likely to result from differences in means.

▶ Example. Students from a number of college departments were given an abstract reasoning test before beginning an introductory course in statistics. This test grades them as 'high' or 'low' in reasoning ability. It was expected that students high on the scale would perform better in the end of course examinations. It is possible, however, that students in different departments might work with different levels of motivation. To avoid this complication, matched sets of students were used. Each set contained students from the same department but contained 3 high and 2 low in abstract reasoning ability. Six sets of students were used. Their examination results were given on a 10 point scale (0 = lower 10 per cent of marks, 9 = top 10 per cent).

SCORES: RELATED; SETS MATCHED ACROSS ABILITY GROUPS
VARIABLES: (a) ABILITY GROUPS, NOMINAL, 2 CATEGORIES
            (b) EXAMINATION RESULTS; RANKABLE SCORES
DATA SUMMARY TABLE:

| | Groups | |
Matched sets	Low	High
1	4, 2	8, 6, 6
2	7, 4	9, 6, 5
3	7, 5	9, 6, 2
4	5, 4	6, 6, 5
5	7, 4	8, 8, 4
6	6, 2	7, 6, 1

$$n_1 = 2$$
$$n_2 = 3$$
$$m = 6$$

1. Rank within sets

| | Groups | | | | |
Set	I (Low)		II (High)		
1	2	1	5	3·5	3·5
2	4	1	5	3	2
3	4	2	5	3	1
4	2·5	1	4·5	4·5	2·5
5	3	1·5	4·5	4·5	1·5
6	3·5	2	5	3·5	1

2. Sum the ranks in group I

$$R = 27.5$$

3.

$$Z = \frac{6 \times 2 \times 6 - 2 \times 27.5}{\sqrt{(\frac{1}{3} \times 6 \times 3 \times 2 \times 6)}} = \frac{72 - 55}{\sqrt{72}} = 2.00$$

Our value of R is significant at better than the 2·5 per cent level using a 1-TAIL test. We use a 1-TAIL test because the overall direction of the result was predicted.

Conclusion. Students who score well on the abstract reasoning test perform better in the end of course examinations.

6.7 Related *t* test

SCORES: RELATED (PAIRS) MATCHED ACROSS VARIABLE a
VARIABLES: (a) NOMINAL; 2 CATEGORIES (GROUPS I AND II)
            (b) NORMALLY DISTRIBUTED SCORES

DATA SUMMARY TABLE:

Notes

X_1, X_{II} are normally distributed scores.

D is the difference score for a matched pair $(X_1 - X_{II})$.

n is the number of matched *pairs*.

Matched pairs	Groups I	II	D (Difference)
1	X_1	X_{II}	D_1
2	X_1	X_{II}	D_2
⋮	⋮	⋮	⋮
n	X_1	X_{II}	D_n

Procedure

1. Compute ΣD, the sum of the difference values.
2. Compute ΣD^2, the sum of the squares of the difference values.
3. Compute t, where

$$t = \frac{\Sigma D}{\sqrt{\left(\dfrac{n\Sigma D^2 - (\Sigma D)^2}{n-1} \right)}} \tag{6.7}$$

4. t has $(n-1)$ degrees of freedom and may be assessed by reference to table 2.7.5. Some critical values have been extracted and are given below. A 1-TAIL test is appropriate when the direction of the difference between the means has been correctly predicted.

2-TAIL	10%	5%	2%	1%
1-TAIL	5%	2·5%	1%	0·1%
$df = 6$	$t \geq 1{\cdot}94$	2·45	3·14	5·21
8	1·86	2·31	2·90	4·50
10	1·81	2·23	2·76	4·14
12	1·78	2·18	2·68	3·93
15	1·75	2·13	2·60	3·73
20	1·72	2·09	2·53	3·55
30	1·70	2·04	2·46	3·39

Inferences. Our two groups can be treated as samples from two populations. We wish to know whether these populations have different means. t is a measure of the difference in the sample means. Table 2.7.5 describes the situation where the following assumptions are met:

(i) The two populations have the same means (i.e., the difference scores are drawn from a population whose mean is zero).

(ii) The scores in both populations are normally distributed. (More strictly this assumption concerns the normality of the population of possible values of D.)

(iii) The samples are drawn at random within the limitations of the need to match the pairs.

A significantly large value of t indicates that one of these assumptions is untrue. The test should not be used when assumption (ii) is not valid. Proper experimental procedure should guarantee assumption (iii). This leaves the implication that the means are different.

▶ Example.   This example compares the examination results of students of two levels of reasoning ability. To control for motivational differences between departments, our sample of 14 students was divided into 7 pairs. Both members of each pair belonged to the same department but one had a high the other a low score on an abstract reasoning test. The examination results were given as a mark from 1–100. Do the results given below support our hypothesis that students having high abstract reasoning ability will perform better in the end of course examinations?

SCORES: RELATED; MATCHED PAIRS
VARIABLES: (a) REASONING ABILITY; NOMINAL, 2 GROUPS
                   (b) EXAMINATION RESULTS; POSSIBLY NORMALLY DISTRIBUTED
                        SCORES
DATA SUMMARY TABLE:

| | Ability groups | | |
Pairs	High	Low	D
1	78	63	+15
2	56	58	−2
3	64	32	+32
4	71	54	+17
5	62	64	−2
6	59	43	+16
7	62	62	0

Notes
$n = 7$.

1. Compute $\Sigma D = 15 - 2 + 32 + 17 - 2 + 16 + 0 = 76$.
2. Compute $\Sigma D^2 = 1802$.
3. Calculate t

$$t = \frac{\Sigma D}{\sqrt{\left(\dfrac{n\Sigma D^2 - (\Sigma D)^2}{n-1}\right)}} = \frac{76}{\sqrt{\left(\dfrac{7 \times 1802 - 76^2}{6}\right)}} = 2\cdot25$$

4. t has $k - 1 = 7 - 1 = 6$ degrees of freedom. Our value of t is significant at better than the 5 per cent level ($t = 1\cdot94$ for 6 df at 5 per cent level, 1-TAIL test).

Conclusion. Students with high abstract reasoning ability perform better in the end of course statistics examinations.

6.8 Friedman's test

SCORES;   RELATED; SETS MATCHED ACROSS VARIABLE a
VARIABLES: (a) NOMINAL; 3 OR MORE CATEGORIES (GROUPS I, II to k)
                   (b) RANKABLE SCORES (X)
DATA SUMMARY TABLE:

	Matched sets	Groups			
		I	II	...	k
	1	X_1	X_1	...	X_1
	2	X_2	X_2	...	X_2
	.	.	.	.	.
	.	.	.	.	.
	.	.	.	.	.
	n	X_n	X_n	...	X_n

Notes
X are rankable scores.
n is the number of matched *sets* of scores.
k is the number of groups.

Procedure
1. For each matched set, rank the scores *within the set* from 1 to k.
2. Calculate $R_1, R_2, \ldots, R_k$, the sum of the ranks in each group.
3. Calculate H, the sum of the squares of the rank totals $H = \Sigma R^2$.
4. The significance of H may be assessed by reference to table 6.8. H must be *larger* than the value given in the table to be significant.
5. Where table 6.8 is inappropriate (e.g., for $k > 3$)

$$\chi^2 = \frac{12H}{nk(k+1)} - 3n(k+1) \tag{6.8}$$

χ^2 has $k - 1$ degrees of freedom and may be assessed by reference to table 2.7.4. Some critical values have been extracted and are given below:

Significance level		5%	2.5%	1%	0.1%
Number of groups	df				
3	2	$\chi^2 \geq 6.0$	7.4	9.2	13.8
4	3	7.8	9.3	11.3	16.3
5	4	9.5	11.1	13.3	18.5

Inferences.   Our k groups of scores may be regarded as samples from k populations. We wish to know if the population means are different from each other. Table 6.8 and the chi square approximation describe the situation where the following assumptions are met:

 (i) The populations have the same distribution characteristics (means, variances, skews, etc.).
(ii) The samples are drawn at random from their respective populations within the limitations of the need to match sets of scores.

A significantly large value of H indicates that one of these assumptions is not true. Assumption (ii) should be guaranteed by good experimental procedure. This leaves the implication that the distribution characteristics are not identical. A large value of H is possibly caused by differences in the medians and, most likely, the means of the populations.

▶ Example.   It was suspected that students from different departments worked with different levels of motivation and diligence in an introductory statistics course. This hypothesis was tested by comparing the end of course examination results for four departments. To compensate for different levels of ability which may exist between departments, the students were grouped into sets of four (1 student per department) where all four were matched according to their results on an abstract reasoning test. The end of course examination result was given on a 10 point scale where 10 points indicates a result in the top 10 per cent of the class, 9 points indicates a result in the next 10 per cent of the class, etc.

Table 6.8 Critical values of H (for $k = 3$) in the Friedman test.

2 - TAIL		10%	5%	2%	0·2%
1 - TAIL		5%	2·5%	1%	0·1%
$N = 3$	$H \geq$	126			
4		215	216	216	
5		332	338	342	350
6		474	482	486	504
7		638	642	650	674
8		818	830	840	866
9		1028	1044	1058	1086

SCORES: RELATED; SETS MATCHED ACROSS DEPARTMENTS
VARIABLES: (a) DEPARTMENT; NOMINAL, 4 GROUPS
(b) EXAMINATION RESULT; RANKABLE
DATA SUMMARY TABLE:

Groups

Set	Biology	Economics	Psychology	Sociology
1	6	4	3	3
2	8	7	7	9
3	7	3	4	7
4	7	1	3	1
5	6	2	7	4

Notes
$n = 5$; $k = 4$.

1. Rank the scores within sets.
2. Calculate $R_1 R_2 R_3 R_4$.

	Biology	Economics	Psychology	Sociology
1	4	3	1·5	1·5
2	3	1·5	1·5	4
3	3·5	1	2	3·5
4	4	1·5	3	1·5
5	3	1	4	2
	17·5	8	12	12·5

3. Calculate H

$$H = \Sigma R^2 = 670 \cdot 5$$

4. For $k > 3$ use formula (6.8)

$$\chi^2 = \frac{12H}{nk(k + 1)} - 3n(k + 1)$$

$$= \frac{12 \times 670 \cdot 5}{5 \times 4 \times 5} - 3 \times 5 \times 5 = 5 \cdot 46$$

$$df = k - 1 = 4 - 1 = 3$$

Our value of χ^2 is not significant ($\chi^2 = 7 \cdot 8$ is required at 5 per cent level).

Conclusion. The evidence is not strong enough to support the hypothesis that students in different departments work with different levels of motivation and diligence.

6.9 Page's L (Trend) test

SCORES: RELATED SETS MATCHED ACROSS VARIABLE a
VARIABLES: (a) RANKABLE CATEGORIES (GROUPS I, II to k)
            (b) RANKABLE SCORES
DATA SUMMARY TABLE:

	Groups			
	I	II	...	K
Matched sets				
1	x_1	x_1	...	x_1
2	x_2	x_2	...	x_2
$\vdots$	$\vdots$	$\vdots$	...	$\vdots$
n	x_n	x_n	...	x_n
	$y_1 = 1$	$y_2 = 2$		$y_k = k$

Notes

x are rankable scores.
n is the number of matched sets of scores.
k is the numbers of groups.
y is the number of each group.

If the trend is descending rather than ascending, reverse the numbering of the groups $(k, \ldots, 2, 1)$.

Procedure

1. For each matched set, rank the scores within the set from 1 to k.
2. Calculate $R_1, R_2, \ldots, R_k$, the sum of the ranks in each group.
3. Calculate L

$$L = \Sigma yR \tag{6.9.1}$$

4. The significance of L may now be assessed by reference to table 6.9. L must be *larger* than the value given in the table to be significant.
5. For large samples, compute:

$$Z = \frac{12L - 3nk(k + 1)^2}{k\sqrt{(n(k^2 - 1)(k + 1))}} \tag{6.9.2}$$

Z is normally distributed and its significance may be assessed by reference to table 2.7.2. Some critical values of Z have been extracted and are given below:

2-TAIL	10%	5%	2%	0·2%
1-TAIL	5%	2·5%	1%	0·1%
$Z \geq$	1·64	1·96	2·33	3·09

Page's L is a test of trend and a 1-TAIL test is appropriate when the observed trend is in the predicted direction.

Inferences. Our K groups of scores may be regarded as samples from K different populations. We wish to know whether the means of these populations are ordered with respect to one another in a specific pattern. L is a measure of the ordering of the samples from group I to group K. A large value of L reflects a strong trend. Table 6.9 and the approximation to the normal distribution describe the situation where the following assumptions are met:

(i) The populations have the same means.
(ii) The samples are drawn at random from their respective populations within the limitations of the need to match sets of scores.

A significantly large value of L indicates that one of these assumptions is not true. Assumption (ii) should be guaranteed by good experimental procedure. This leaves the implication that the means of the K groups are not the same. In particular, it suggests that the means are ordered to form a trend.

Table 6.9   Critical values of L in Page's trend test.

	2-TAIL 1-TAIL	10% 5%	5% 2·5%	2% 1%	0·2% 0·1%
$k = 3$	$n = 2$	$L \geq 28$			
	3	41	42	42	
	4	54	55	55	56
	5	67	68	68	70
	6	79	80	81	83
$k = 4$	$n = 2$	58	59	60	
	3	84	86	87	89
	4	111	112	114	117
	5	137	138	140	145

▶ Example.   This example tests the relationship between abstract reasoning ability and end of course examination results. Scores on the abstract reasoning test were divided into 5 grades (1, lowest 20 per cent; 5, highest 20 per cent). From each department, 5 students were selected, one from each reasoning category. The end of course examination results were given on a 10 point scale. It was predicted that the examination results would show a trend across the ability groups. Do the results support this prediction?

SCORES: RELATED, SETS MATCHED BY DEPARTMENT ACROSS ABILITY LEVELS.
VARIABLES: (a) ABILITY LEVEL; RANKED CATEGORIES
                      (b) EXAMINATION RESULTS; RANKABLE SCORES
DATA SUMMARY TABLE:

		Ability groups				
Matched sets		1	2	3	4	5
1. Zoology		9	2	7	3	10
2. Economics		1	1	10	7	8
3. Psychology		6	1	7	6	9
4. Sociology		4	8	2	9	8
5. Education		6	3	3	7	7
6. Botany		8	10	8	8	6

1. Rank the scores within each set.
2. Calculate R_1, R_2, R_3, R_4, R_5.

Matched sets	1	2	3	4	5
1	4	1	3	2	5
2	1·5	1·5	5	3	4
3	2·5	1	4	2·5	5
4	2	3·5	1	5	3·5
5	3	1·5	1·5	4·5	4·5
6	3	5	3	3	1
R	16	13·5	17·5	20	23
y	1	2	3	4	5
yR	16	27	52·5	80	115 $\Sigma yR = 290\cdot5$

3. The value of L is 290·5.
4. Compute Z

$$Z = \frac{12L - 3nk(k+1)^2}{k\sqrt{n(k^2-1)(k+1)}} = \frac{12 \times 290\cdot5 - 3 \times 6 \times 5 \times 36}{5\sqrt{6 \times 24 \times 6}}$$

$$= \frac{246}{5\sqrt{864}} = 1\cdot67$$

Our value of Z is significant at the 5 per cent level.

A 1-TAIL test has been used because we expected that examination performance would improve with ability levels.

Conclusion.   Our results were significant at the 5 per cent level on a 1-TAIL test and confirm our prediction that students with better reasoning ability would perform better in the end of course examination.

7. Analysis of variance

7.1 Comparing variance estimates

7.1.1 Comparing two variance estimates

The F distribution can be used to help decide whether two samples were drawn from the same population. Compute F:

$$F_{n_1-1,n_2-1} = \frac{\text{var. est. (1)}}{\text{var. est. (2)}} \qquad (7.1.1)$$

where var. est. (1) and var. est. (2) are, respectively, estimates of the population variance based on the two samples.

F has associated with it $(n_1 - 1)$ and $(n_2 - 1)$ degrees of freedom.

n_1 and n_2 are the sample sizes.

Table 2.7.3 can now be used to assess the significance of F. This table is based on situations where the following assumptions are met:

(i) The two samples are drawn from populations with the same variance.
(ii) The samples are drawn at random.
(iii) The populations from which the samples are drawn are normally distributed.

Significantly large values indicate that at least one of these assumptions may be untrue. If assumptions (ii) and (iii) can be guaranteed we may tentatively conclude that our two samples were drawn from populations with different variances.

1- and 2-TAIL tests. Table 2.7.3 gives values for a 1-TAIL test where a prediction is made (before the results are available) as to which variance estimate will be larger. It is the 'predicted larger' estimate which is used as the numerator.

$$F_{n_1-1,n_2-1} = \frac{\text{Predicted larger var. est. (1)}}{\text{Predicted smaller var. est. (2)}}$$

When no prediction has been made we may calculate F in one of two different ways:

$$F_{n_1-1,n_2-1} = \frac{\text{var. est. (1)}}{\text{var. est. (2)}} \quad \text{or} \quad F_{n_2-1,n_1-1} = \frac{\text{var. est. (2)}}{\text{var. est. (1)}}$$

and choose the larger of the two. Accordingly, we must *double* the percentage significance value and this becomes a 2-TAIL test.

▶ Example.   It is suspected that women will prove more variable in their response to examination conditions. A sample of 25 examination results was taken which included 12 male and 13 female scores. Are the women's scores more variable?

Men's scores:

$$39, 45, 49, 53, 57, 60, 60, 61, 65, 71, 78, 82$$

$$N = 12$$

$$\sigma^2 = \frac{\Sigma X^2 - (\Sigma X)^2/N}{(N-1)} = \frac{45\,000 - 720^2/12}{11} = 163 \cdot 6 \ (11 \ df)$$

Women's scores:

$$24, 28, 31, 37, 53, 59, 64, 70, 75, 75, 83, 90, 91$$

$$N = 13$$

$$\sigma^2 = \frac{53\,496 - 780^2/13}{12} = 558 \ (12 \ df)$$

$$F_{12,11} = \frac{558}{163 \cdot 6} = 3 \cdot 4$$

Our value of F is significant at the 5 per cent level, we may infer that women are more variable in their examination results than men. (When appropriate entries for df are not present in the table use the next *smaller* values. In this case, look for the critical values corresponding to $F_{10,10}$.)

7.1.2 Homogeneity of variance test

Procedure. When we have three or more samples, we may wish to test whether all groups were drawn from populations with the same variances. Equality of group variances is an important assumption in many analysis of variance designs (e.g., 7.3 assumption iii). A simple approximate test is made possible by the F_{max} statistic:

$$F_{max(df_1, df_2)} = \frac{\text{Largest of the group variance estimates}}{\text{Smallest of the group variance estimates}} \qquad (7.1.2a)$$

where df_1, df_2 are the degrees of freedom associated with the two variance estimates.

Inferences. F_{max} may be assessed using table 7.1.2 which gives critical values at the 5 and 1 per cent significance levels only. In this table, k is the number of groups and df refers to the *average* value of df for the samples. Obtained values of F_{max} must be greater than or equal to the values given in the table to be significant.

The F_{max} table applies when the following assumptions are true:

(i) The samples were drawn from populations which have the same variances.
(ii) The samples were drawn from normally distributed populations.

A significantly large value of F_{max} may indicate that at least one of these two assumptions is untrue. The test should only be used when assumption (ii) is valid and then we may conclude that the samples are not all drawn from populations with the same variances.

▶ **Example.** The following scores are ready for analysis by one way analysis of variance. An homogeneity of variance test is required before undertaking the analysis.

	Group A	Group B	Group C
	63	43	53
$n = 5$	47	57	61
$df = 4$	67	59	41
	81	46	43
ΣX	258	205	198
ΣX^2	17 228	10 695	10 060
df	3	3	3
var. est.	195·667	62·917	86·33

$$F_{max} = \frac{\text{Largest var. est.}}{\text{Smallest var. est.}} = \frac{195·667}{62·917} = 3·1$$

$$k = 3$$

$$df = 3$$

It is clear from table 7.1.2 that our value of F_{max} is too small to be significant even at the 5 per cent level. Accordingly, we are unable to reject the idea that the samples were drawn from populations with the same variances.

Table 7.1.2 Critical values of F_{max}.

Significance level			5%	1%
Number of estimates $k = 2$	Degrees of freedom $df = 2$	$F \geq$	39	199
	3		15·4	47·5
	4		9·6	23·2
	5		7·2	14·9
	10		3·7	5·9
	30		2·1	2·6
	60		1·7	2·0
$k = 3$	$df = 2$		87·5	448
	3		27·8	85
	4		15·5	37
	5		10·8	22
	10		4·9	7·4
	30		2·4	3·0
	60		1·9	2·2
$k = 4$	$df = 2$		142	729
	3		39·2	120
	4		20·6	49
	5		13·7	28
	10		5·7	8·6
	30		2·6	3·3
	60		2·0	2·3
$k = 5$	$df = 2$		202	1036
	3		50·7	151
	4		25·2	59
	5		16·3	33
	10		6·3	9·6
	30		2·8	3·4
	60		2·0	2·4
$k = 6$	$df = 2$		266	1362
	3		62·0	184
	4		29·5	69
	5		18·7	38
	10		6·9	10·4
	30		2·9	3·6
	60		2·4	2·4
$k = 8$	$df = 2$		403	2063
	3		83·5	249
	4		37·5	89
	5		22·9	46
	10		7·9	11·8
	30		3·1	3·8
	60		2·2	2·5
$k = 10$	$df = 2$		550	2813
	3		104	310
	4		44·6	106
	5		26·5	54
	10		8·7	12·9
	30		3·3	4·0
	60		2·3	2·6

This table is abridged from Table 31 in *Biometrika Table for Statisticians*, vol 1 (2nd ed.) Cambridge University Press, 1958 Edited by E. S. Pearson and H. O. Hartley.

7.2 Analysis of variance (General)

Analysis of variance also allows us to compare group means by calculating a variance estimate based on the variability between the group means. This variance estimate is then compared with another variance estimate based on the variability of scores within the groups. Procedures given in section 7.1 are used for this latter comparison:

$$F = \frac{\text{Between group variance estimate}}{\text{Within group variance estimate}}$$

When the groups are samples drawn from populations with the same means, the value of F will usually be non-significant. Significant values of F suggest that the groups are samples drawn from populations with different means. This test is only accurate when the various populations from which the samples are drawn have the same variance.

7.2.1 Terminology

A special language has grown up around analysis of variance. Some of these terms are explained below by reference to the usual computational formula for a variance estimate based on a sample.

$$\text{var. est.} = \frac{\Sigma X^2 - \dfrac{(\Sigma X)^2}{N}}{N - 1}$$

In analysis of variance, parts of this formula are calculated separately and brought together in the 'Summary Table'.

$\dfrac{(\Sigma X)^2}{N}$ is called the *correction term* or C

$\Sigma X^2 - \dfrac{(\Sigma X)^2}{N}$ is called the *sum of squares, S.S.*

it can be written $\Sigma X^2 - C$

$N - 1$ are the *degrees of freedom, df*

var. est.   is often called the *mean square (M.S.)*

Our variance estimate can, therefore, be written in many ways

$$\text{var. est.} = \text{Mean square} = \text{M.S.}$$

$$= \frac{SS}{df} = \frac{\Sigma X^2 - C}{df} = \frac{\Sigma X^2 - C}{N - 1}$$

They all say the same thing.

When basing our variance estimate on *group means* (rather than individual scores) we need a little additional notation. The formulae use group totals (t) rather than group means for convenience

t group total ($t_1, t_2, \ldots, t_k$)
k number of groups
n_t number of scores in each group, i.e., the number of scores which contribute to the *total* (t) in question
T the grand total (Σt)
N the total number of scores (Σn_t).

We calculate:
1. Correction term

$$C = \frac{T^2}{N}$$

2. Sum of squares

$$S.S. = \sum \frac{t^2}{n_t} - C$$

Notice that each group total is squared and then divided by the number of scores (n_t) which have *contributed to that particular total*.
3. Degrees of freedom $df = k - 1$
usually the degrees of freedom are 1 less than the *number of totals* used in the sum of squares.
4. Variance estimate (mean square)

$$\text{var. est.} = \frac{SS}{df} = M.S.$$

7.2.2 *F* ratios

In a simple 1 Factor (1 way) analysis of variance the *F* ratio given above is used. A more general expression is:

$$F = \frac{\text{Main effect variance estimate}}{\text{Error variance estimate}}$$

In many cases, the error variance estimate is merely the within groups or within cells variance estimate. This is not always the case, however, and often there is no within cells estimate. The choice of error term is discussed further in the section on *fixed* and *random* effects.

7.2.3 1- and 2-TAIL tests

Normally, these considerations do not apply to analysis of variance and it is sufficient to quote the percentage significance level given at the head of table 2.7.3 under 1-TAIL. They can apply, however, when the numerator of the *F* ratio has *only* 1 *degree of freedom*. In this case, one can predict one of two possibilities before the analysis, for example whether one of two groups is larger or smaller than the other or whether a linear trend has positive or negative slope. In this case, the percentage significance level must be *halved* again. Thus, on a 1-TAIL test, an *F* ratio of 4·3 with 1 and 20 *df* is significant at the 2·5 per cent significance level.

7.2.4 Fixed and random effects

Each variable under study is represented by at least two values or levels. When these levels exhaust all the possibilities of interest they are called 'fixed'. *Sex* is a fixed effect because Male and Female exhaust all the possibilities. *Age* will also be a fixed effect if the categories used cover the full range of interest. Some variables, however, constitute only a sample of the total range these are 'random' effects. For example, *subjects* can be a random effect when they merely constitute a sample.

Some variables are not so easy to classify but depend on the researcher's intentions. For example, the choice of three departments, Botany, Psychology, Education could constitute either a fixed or random effect. If the intention is to generalize to all possible departments, then the effect is said to be random. If these three departments represent the full range of interest then the effects are fixed.

The fixed vs. random effects distinction is important in the choice of proper error term. Table 7.2.4 illustrates how the choice of error term is affected by the presence of random effects in even straightforward factorial designs. Table 7.2.4 should be used for reference. A simple rule is that the proper error term for any effect is the 'within cells variance estimate' *except* where random effects are present. When these are present, the proper error for any effect is the interaction of that effect with the random effect.

The situation is complicated further by the absence of replications, that is when there is only one score in each cell. Often, this means that there is no proper error term. Where a proper error term can be found this is given. Where none exist, a substitute is suggested (within brackets). Use of these substitutes often results in low power, conservative tests which make significance difficult to achieve.

Table 7.2.4 Choice of appropriate error term in factorial analysis of variance designs with fixed and random effects, with and without 'within cells variance estimates'. Brackets indicate that no exact error term is available.

	Effect	Type	Error term	No within cells variance estimate
1 Factor (1 Way)	A	fixed	within cells	—
2 Factor (2 Way)	A	fixed	within cells	$(A \times B)$
	B	fixed	within cells	$(A \times B)$
	$A \times B$		within cells	—
	A	fixed	$A \times B$	$A \times B$
	B	random	within cells	$(A \times B)$
	$A \times B$		within cells	
3 Factor (3 Way)	A	fixed	within cells	$(A \times B \times C)$
	B	fixed	within cells	$(A \times B \times C)$
	C	fixed	within cells	$(A \times B \times C)$
	$A \times B$		within cells	$(A \times B \times C)$
	$A \times C$		within cells	$(A \times B \times C)$
	$B \times C$		within cells	$(A \times B \times C)$
	$A \times B \times C$		within cells	—
	A	fixed	$A \times C$	$A \times C$
	B	fixed	$B \times C$	$B \times C$
	C	random	within cells	$(A \times B \times C)$
	$A \times B$		$A \times B \times C$	$A \times B \times C$
	$A \times C$		within cells	$(A \times B \times C)$
	$B \times C$		within cells	$(A \times B \times C)$
	$A \times B \times C$		within cells	—
	A	fixed	$A \times B$ or $A \times C$	$A \times B$ or $A \times C$
	B	random	$B \times C$	$B \times C$
	C	random	$B \times C$	$B \times C$
	$A \times B$		$A \times B \times C$	$A \times B \times C$
	$A \times C$		$A \times B \times C$	$A \times B \times C$
	$B \times C$		within cells	$(A \times B \times C)$
	$A \times B \times C$		within cells	—

129

7.3 Factorial designs

An analysis of variance design is said to be factorial when at least one score exists for each combination of all levels of the variables. Thus, in a 2×4 design there must be at least 8 scores. When each combination is repeated 2 or more times the design becomes *factorial with replications*. The replications form the basis of the within cells variance estimate.

Three examples are given (1, 2 and 3 variable analysis) which should provide useful working models. You are warned that analysis of variance usually contains special assumptions associated with each particular design. These are dealt with at length in advanced texts. Tables of critical values of F apply when the following assumptions are true:

  (i) The samples are taken from normally distributed populations.

 (ii) The samples are made at random.

(iii) The variances of the sampled populations are equal.

(iv) The means of the sampled populations are equal.

Analysis of variance should not be used when the populations under study are not normally distributed or when the samples are not made at random. Under these circumstances, a significantly large F ratio indicates that either assumption (iii) or (iv) is untrue. It is not always easy to check assumption (iii) but it should be borne in mind when concluding that the population means are different.

Examples 7.3.1, 7.3.2 and 7.3.3. In these three examples we have isolated three groups of students of 'high', 'medium' and 'low' abstract reasoning ability. We wish to know whether reasoning ability influences their end of course examination mark. In fact 'high', 'medium' and 'low' are rankable categories and would be suitable for further analysis using trend tests.

In example 1 the three groups are chosen at random from all available students on the course.

In example 2 we have two students in each group from the same department.

In example 3 we have taken the opportunity to study two types of examination conditions. Under the first condition (short), students must complete the examination in three hours. Under the second condition (long) students have up to six hours to complete the same work. Male and female students constitute subdivisions of the three reasoning ability classes. This is a 3 variable analysis of variance, (a) examination type, (b) sex and (c) reasoning ability.

7.3.1 One way example

Reasoning ability (I.Q.)	High	Medium	Low	
	63	43	53	$k = 3$
	47	57	61	$n_1 = 4$
	67	59	41	$n_2 = 5$
	81	46	43	$n_3 = 4$
		71		$N = 13$
	$t_1 = 258$	$t_2 = 276$	$t_3 = 198$	$T = 732$
				$\Sigma X^2 = 43\,024$

1. Correction term:

$$C = T^2/N = 732^2/13 = 41\,217 \cdot 2$$

2. Total sum of squares:

$$\text{T.S.S.} = \Sigma X^2 - C = 63^2 + 47^2 + \ldots + 43^2 - C = 43\,024 - 41\,217 \cdot 2 = 1806 \cdot 8$$

Total degrees of freedom:

$$T \cdot df = N - 1 = 13 - 1 = 12$$

3. I.Q. sum of squares:

$$\text{I.Q.S.S.} = \sum \frac{t^2}{n_t} - C = \frac{258^2}{4} + \frac{276^2}{5} + \frac{198^2}{4} - C$$

$$= 41\,677 \cdot 2 - 41\,217 \cdot 2 = 460$$

I.Q. degrees of freedom:

$$\text{I.Q.}df. = k - 1 = 3 - 1 = 2$$

$$\text{I.Q. variance estimate} = \text{I.Q.S.S.}/\text{I.Q.}df = 460/2 = 230$$

4. Within groups sum of squares:

$$\text{W.G.S.S.} = \text{Total S.S.} - \text{I.Q.S.S.}$$

$$= 1806 \cdot 8 - 460 = 1346 \cdot 8$$

Within groups degrees of freedom:

$$\text{W.G.}df = \text{Total } df - \text{I.Q.}df.$$

$$= 12 - 2 = 10$$

Within groups variance estimate:

$$\text{W.G. var. est.} = \text{W.G.S.S.}/\text{W.G.}df = 1346 \cdot 8/10 = 134 \cdot 68$$

5. F (I.Q.) ratio:

$$F = \text{I.Q. var. est.}/\text{W.G. var. est.} = 230/134 \cdot 68 = 1 \cdot 708$$

$$df = \text{I.Q.}df, \text{W.G.}df = 2, 10$$

A value of $4 \cdot 1$ is required at 5 per cent significance level for 2 and 10 df. Our value is not significant.

6. *Summary table*

Source	S.S.	df	Var. est.	F	df	Significance
I.Q.	460	2	230	$1 \cdot 7$	2, 10	NS
Within groups	1346·8	10	134·68			
Total	1806·8	12				

Conclusion. The data do not indicate that reasoning ability is related to end of course statistics examination results.

7.3.2 Two way example

reasoning ability (I.Q.)

Department	High	Medium	Low	Totals
Zoology	73, 61	61, 64	63, 41	363
Psychology	64, 72	60, 61	51, 64	372
Sociology	71, 63	70, 68	47, 48	367
Education	83, 69	46, 67	53, 62	380
	556	497	429	1482

Totals $N = 24$; $T = 1482$;

1. Correction term:

$$C = T^2/N = 1482^2/24 = 91\,513 \cdot 5$$

2. Total sum of squares:

$$\text{T.S.S.} = \Sigma X^2 - C = 73^2 + 61^2 + \ldots + 53^2 + 62^2 - C = 93\,770 - 91\,513 \cdot 5 = 2256 \cdot 5$$

Total $df = N - 1 = 24 - 1 = 23$

3. Department sum of squares:

$$\text{D.S.S.} = \frac{\Sigma t^2}{n_t} - C = \frac{363^2 + 372^2 + 367^2 + 380^2}{6} - C$$

$$= 91\,540 \cdot 33 - 91\,513 \cdot 5 = 26 \cdot 83$$

Department degrees of freedom:

$$\text{D}.df = \text{Number of groups} - 1 = 4 - 1 = 3$$

Department variance estimate:

$$\text{Dept. var. est.} = \text{D.S.S.}/\text{D}.df = 26 \cdot 83/3 = 8 \cdot 94$$

4. I.Q. sum of squares:

$$\text{I.Q.S.S.} = \frac{\Sigma t^2}{n_t} - C = \frac{556^2 + 497^2 + 429^2}{8} - C$$

$$= 92\,523 \cdot 25 - 91\,513 \cdot 5 = 1009 \cdot 75$$

I.Q. degrees of freedom:

$$\text{I.Q}.df = \text{Number of groups} - 1 = 3 - 1 = 2$$

I.Q. variance estimate:

$$\text{I.Q. var. est.} = \text{I.Q.S.S.}/\text{I.Q}.df = 1009 \cdot 75/2 = 504 \cdot 87$$

132

5. I.Q. × Department interaction effect:

		High	Medium	Low
Totals	Z	134	125	104
	P	136	121	115
	S	134	138	95
	E	152	113	115

I.Q. × Dept. S.S. $= \dfrac{\Sigma t^2}{n_t} - \text{I.Q.S.S.} - \text{D.S.S.} - C$

$$= \frac{134^2 + 136^2 + \ldots + 115^2}{2} - 1009.75 - 26.83 - 91\,513.5$$

$$= 390.9$$

I.Q. × Dept. df = I.Q.df × D.df = 2 × 3 = 6

I.Q. × Dept. var. est. = S.S./df = 390.9/6 = 65.15

6. Within cells sum of squares:

$$\text{W.C.S.S.} = \text{T.S.S.} - \text{All other S.S.}$$

$$= 2256.5 - 1009.75 - 26.83 - 390.9 = 829.0$$

$$\text{W.C. } df = \text{T. } df - \text{All other } df$$
$$= 23 - 3 - 2 - 6 = 12$$

Within cells variance estimate:

$$\text{W.C. var. est.} = \text{S.S.}/df = 829.0/12 = 69.1$$

7. Summary table

Source	S.S.	df	var. est.	F	df	Significance
Department	26.83	3	8.94	0.1	3, 12	NS
Reasoning (I.Q.)	1009.75	2	504.87	7.3	2, 12	< 2.5%
I.Q. × Dept.	390.9	6	65.15	0.9	6, 12	NS
*Within cells	829.0	12	69.1			
Total	2256.5	23				

* Within cells was used here as the error term.

Conclusion. The data provide support for the idea that reasoning ability influences examination performance. No department effect or reasoning ability by department interaction effect was found.

7.3.3 Three way example

Reasoning ability	High				Medium				Low			
Sex	Male		Female		Male		Female		Male		Female	
Examination	Short	Long	S	L	S	L	S	L	S	L	S	L
	68	72	72	65	58	59	55	65	43	55	37	61
	70	81	70	87	60	75	75	71	56	55	54	57
	138	153	142	152	118	134	130	136	99	110	91	118

$$N = 2 \times 2 \times 3 \times 2 = 24; \qquad T = 1521$$

1. Correction term: $C = T^2/N = 1521^2/24 = 96\,393\cdot4$.

2. Total:

$$\text{S.S.} = \Sigma X^2 - C = 99\,423 - 96\,393\cdot4 = 3029\cdot6$$

$$df = N - 1 = 24 - 1 = 23$$

3. Reasoning ability (I.Q.):

$$\text{Totals: High} = 585, \text{Medium} = 518, \text{Low} = 418: n_t = 8$$

$$\text{I.Q.S.S.} = \frac{\Sigma t^2}{n_t} - C = \frac{585^2 + 518^2 + 418^2}{8} - 96\,393\cdot4 = 1765\cdot7$$

$$\text{I.Q. } df = \text{Number of groups} - 1 = 3 - 1 = 2$$

$$\text{I.Q. var. est.} = \text{S.S.}/df = 1765\cdot7/2 = 882\cdot8$$

4. Sex:

$$\text{Totals: Male} = 752, \text{Female} = 769: n_t = 12$$

$$\text{Sex S.S.} = \frac{\Sigma t^2}{n_t} - C = \frac{752^2 + 769^2}{12} - 96\,393\cdot4 = 12\cdot0$$

$$\text{Sex } df = \text{Number of groups} - 1 = 2 - 1 = 1$$

$$\text{Sex var. est.} = \text{S.S.}/df = 12\cdot0/1 = 12\cdot0$$

5. Examination type:

$$\text{Totals: Short} = 718, \text{Long} = 803: n_t = 12$$

$$\text{Exam. S.S.} = \frac{\Sigma t^2}{n_t} - C = \frac{718^2 + 803^2}{12} - 96\,393\cdot4 = 301\cdot0$$

$$\text{Exam. } df = \text{Number of groups} - 1 = 2 - 1 = 1$$

$$\text{Exam. var. est.} = \text{S.S.}/df = 301\cdot0/1 = 301\cdot0$$

6. Reasoning ability (I.Q.) × Sex interaction

Totals

	High	Medium	Low	
Male	291	252	209	
Female	294	266	209	$n_t = 4$

$$\text{S.S.} = \frac{\Sigma t^2}{n_t} - \text{I.Q.S.S.} - \text{Sex S.S.} - C$$

$$= \frac{291^2 + 294^2 + \dots + 209^2}{4} - 1765\cdot7 - 12\cdot0 - 96\,393\cdot4 = 13\cdot65$$

$$df = \text{I.Q. } df \times \text{Sex } df = 2 \times 1 = 2$$

var. est. = S.S./df = 13·65/2 = 6·8

7. Reasoning ability (I.Q.) × Examination type interaction

Totals

	High	Medium	Low	
Short	280	248	190	
Long	305	270	228	$n_t = 4$

$$\text{S.S.} = \frac{\Sigma t^2}{n_t} - \text{I.Q.S.S.} - \text{Exam. S.S.} - C$$

$$= \frac{280^2 + 305^2 + \dots + 228^2}{4} - 1765\cdot7 - 301\cdot0 - 96\,393\cdot4 = 18\cdot15$$

$$df = \text{I.Q. } df \times \text{Exam. } df = 2 \times 1 = 2$$

var. est. = S.S./df = 18·15/2 = 9·1

8. Sex × Examination type interaction

Totals

	Short	Long	
Male	355	397	
Female	363	406	$n_t = 6$

$$\text{S.S.} = \frac{\Sigma t^2}{n_t} - \text{Sex S.S.} - \text{Exam. S.S.} - C$$

$$= \frac{355^2 + 363^2 + 397^2 + 406^2}{6} - 12\cdot0 - 301 - 96\,393\cdot4 = 0\cdot1$$

$$df = \text{Sex } df \times \text{Exam. } df = 1 \times 1 = 1$$

var. est. = S.S./df = 0·1/1 = 0·1

9. Reasoning ability (I.Q.) × Sex × Examination type interaction

Totals: Use column totals from summary table: $n_t = 2$

$$\text{S.S.} = \frac{\Sigma t^2}{n_t} - \text{I.Q.S.S.} - \text{Sex S.S.} - \text{Exam. S.S.} - \text{I.Q.} \times \text{Sex S.S.} - \text{I.Q.} \times \text{Exam. S.S.}$$

$$- \text{Sex} \times \text{Exam. S.S.} - C$$

$$= \frac{138^2 + 153^2 + \ldots + 118^2}{2} - 1765 \cdot 7 - 12 \cdot 0 - 301 \cdot 0 - 13 \cdot 6 - 18 \cdot 15 - 0 \cdot 1 - 96\,393 \cdot 4$$

$$= 47 \cdot 6$$

$$df = \text{I.Q.}\, df \times \text{Sex}\, df \times \text{Exam.}\, df = 2 \times 1 \times 1 = 2$$

var. est. $= \text{S.S.}/df = 47 \cdot 6/2 = 23 \cdot 8$

10. Within cells:

W.C. S.S. $= \text{Total S.S.} - \text{All other S.S.} = 3029 \cdot 6 - 2158 \cdot 1 = 871 \cdot 5$

$$df = \text{Total}\, df - \text{All other}\, df = 23 - 11 = 12$$

var. est. $= \text{S.S.}/df = 871 \cdot 5/12 = 72 \cdot 6$

Summary table

Source	S.S.	df	var. est.	F	df	Significance
Reasoning ability (I.Q.)	1765·7	2	882·8	12·2	2, 12	< 1%
Sex	12·0	1	12·0	0·16	1, 12	NS
Examination type	301·0	1	301·0	4·12	1, 12	NS
I.Q. × Sex	13·6	2	6·8	0·09	2, 12	NS
I.Q. × Examination	18·1	2	9·1	0·12	2, 12	
Sex × Examination	0·1	1	0·1	—		
I.Q. × Sex × Examination	47·6	2	23·8	0·33	2, 12	
*Within cells	871·5	12	72·6			
Total	3029·6	23				

* The within cells has been used as our error variance estimate. This assumes that the three main effects are fixed. (It could be argued that Examination type is either fixed or random.)

Conclusion. Our results indicate that higher reasoning ability produces better examination results. No other effects are significant although it is possible that the longer examination may yield better performances.

7.4 Other designs

Factorial analysis of variance requires that at least one score should be available for every possible combination of each level of each variable. In a $3 \times 3 \times 4$ design there are 36 combinations possible and these will need to be represented at least once (and preferably twice) in a factorial analysis. There are, however, situations where not all conceivable combinations are possible. These have forced the development of special forms of analysis of variance. Two of the more popular of these (Latin square and split plot) are illustrated below.

7.4.1 Latin square design

This design is particularly useful when a number of treatments have to be applied to the same subject or group of subjects. Clearly, the order in which a treatment is presented may influence its effectiveness. A factorial design would require that each subject experienced each treatment first, each treatment second, etc.! This impossible situation can be handled by giving each subject the treatments in a different order. The order is determined by a Latin square which guarantees (a) that each subject receives each treatment once and (b) that each treatment is presented once in each order position.

Here is a Latin square for four treatments ($T1$, $T2$, $T3$, $T4$)

		Order		
Subject	1st	2nd	3rd	4th
Mr A	$T2$	$T4$	$T3$	$T1$
Mr B	$T1$	$T3$	$T4$	$T2$
Mr C	$T3$	$T2$	$T1$	$T4$
Mr D	$T4$	$T1$	$T2$	$T3$

Such squares are readily constructed and can be varied by swapping rows or columns or both at random.

This Latin square design does not permit the computation of interaction terms. Indeed, the design *assumes* that there are *no* significant interactions among order, subjects and treatments. When it is suspected that these may exist, this design should not be used.

This design can also be used where economy is required. The Latin square illustrated above contains only 16 scores and yet it is a $4 \times 4 \times 4$ design. A factorial approach would have required 48 scores. When there are no interactions, then the Latin square design can bring many potentially vast experiments within the realm of small scale laboratory research.

► **Example.** It was decided that an introductory course in statistics would be improved if some film material were introduced. The educational film catalogues offered five short films on introductory statistics. All five were hired and shown on an experimental basis. Students were asked to rate the films for acceptability. To control for order effects, students were split into departmental groups; each department was shown the films in a different order determined by the Latin square below.

Latin square:

Department	Order				
	1st	2nd	3rd	4th	5th
Botany and Zoology	A	D	C	B	E
Psychology	D	B	A	E	C
Sociology	C	A	E	D	B
Education	E	C	B	A	D
Economics	B	E	D	C	A

Films: A, B, C, D, E

Each departmental group produced a rating on a scale from 1–50 for each film. Are the films equally acceptable?

Department	Order					Total
	1st	2nd	3rd	4th	5th	
Botany and Zoology	23	27	33	31	17	131
Psychology	40	26	17	18	28	129
Sociology	39	26	26	37	36	164
Education	38	21	22	23	16	120
Economics	24	20	22	26	20	112
	164	120	120	135	117	656

Film totals

$$\text{Film A} = 23 + 17 + 26 + 23 + 20 = 109$$
$$B = 31 + 26 + 36 + 22 + 24 = 139$$
$$C = 33 + 28 + 39 + 21 + 26 = 147$$
$$D = 27 + 40 + 37 + 16 + 22 = 142$$
$$E = 17 + 18 + 26 + 38 + 20 = 119$$

1. $C = T^2/N = 656^2/25 = 17\,213 \cdot 44$

2. $\text{T.S.S.} = \Sigma X^2 - C = 23^2 + 27^2 + 33^2 + \ldots + 26^2 + 20^2 - C$

   $= 18\,498 - 17\,213 \cdot 44 = 1284 \cdot 56$

   Total $df = N - 1 = 25 - 1 = 24$

3. $\text{Film S.S.} = \dfrac{\Sigma t^2}{n_t} - C = \dfrac{109^2 + 139^2 + 147^2 + 142^2 + 119^2}{5} - C$

   $= 17\,427 \cdot 2 - 17\,213 \cdot 44 = 213 \cdot 76$

   Film $df = k - 1 = 5 - 1 = 4$: var. est. $= \text{S.S.}/df = 213 \cdot 76/4 = 53 \cdot 4$

4. Department S.S. $= \dfrac{\Sigma t^2}{n_t} - C = \dfrac{131^2 + 129^2 + 164^2 + 120^2 + 112^2}{5} - C$

$$= 17\,528 \cdot 4 - 17\,213 \cdot 44 = 315$$

Department $df = k - 1 = 5 - 1 = 4$: var. est. $= $ S.S./$df = 315/4 = 78 \cdot 7$

5. Order S.S. $= \dfrac{\Sigma t^2}{n_t} - C = \dfrac{164^2 + 120^2 + 120^2 + 135^2 + 117^2}{5} - C$

$$= 17\,522 - 17\,213 \cdot 44 = 308 \cdot 56$$

Order $df = k - 1 = 5 - 1 = 4$; var. est. $= $ S.S./$df = 308 \cdot 56/4 = 77 \cdot 1$

6. Residual S.S. $= $ Total S.S. $- $ Film S.S. $- $ Dept. S.S. $- $ Order S.S.

$$= 1284 \cdot 56 - 213 \cdot 76 - 315 - 308 \cdot 56 = 447 \cdot 24$$

Residual $df = $ Total $df - $ Film $df - $ Dept. $df - $ Order df

$$= 24 - 5 - 5 - 5 = 9$$

Residual var. est. $= $ S.S./$df = 447 \cdot 24/9 = 49 \cdot 7$

Summary table:

Source	S.S.	df	var. est.	F	df_1, df_2	Significance
Film	213·76	4	53·4	1·1	4, 9	NS
Department	315	4	78·7	1·6	4, 9	NS
Order	308·56	4	77·1	1·6	4, 9	NS
Residual	447·24	9	49·7			
Total	1284·56	24				

Conclusion. The films do not produce significantly different ratings. It is also of interest to note that there was no order effect nor any significant difference between departments.

7.4.2 Split plot design

When subjects or objects receive all levels of some variables but only one level of other variables, the design is called split plot. Thus a rat may be given, on separate occasions, all of the drugs in the study. The same rat cannot, however, belong to both of the *male* and *female* groups or, say, the *fat* and *thin* groups.

Obviously the requirement of a factorial design—that all subjects should experience all possible combinations of treatments—cannot be met. Split plot analysis is specifically designed to handle this situation. In the example below, we have six departments. Three departments belong to one treatment group and three to the other. We can say that the departments are split into two plots, where each plot receives a different treatment. (The term plot has its origin in agricultural studies.)

▶ Example.   The data in this example came from an experiment to investigate the value of practical statistics classes to supplement the theory lectures. Because of timetabling problems it was not possible to assign students at random to the two groups (a) 'with practice' and (b) 'without practice'. As a result, we were obliged to assign at random, three departments to each group. The analysis, given below, involves the examination results of three students from each department. Because 'ability level' is a complicating factor, the three students were chosen, one from each of three ability levels.

		High	Medium	Low	Total
With	Botany	87	72	68	227
practice	Sociology	79	53	53	185
sessions	Psychology	81	50	61	192
					604
Without	Zoology	88	59	47	194
practice	Education	64	64	46	174
sessions	Economics	71	61	41	173
					541
		470	359	316	1145

$$T = 1145, \qquad N = 18$$

1. $C = T^2/N = 1145^2/18 = 72\,834 \cdot 72$

2. Total S.S. $= \Sigma X^2 - C = 76\,203 - 72\,834 \cdot 72 = 3368 \cdot 28$

   Total $df = N - 1 = 18 - 1 = 17$

3. Departments

   $$\text{Departments S.S.} = \frac{\Sigma t^2}{n_t} - C = \frac{227^2 + 185^2 + 192^2 + 194^2 + 174^2 + 173^2}{3} - C$$

   $$= 73\,486 \cdot 33 - 72\,834 \cdot 72 = 651 \cdot 61$$

   Departments $df = k - 1 = 6 - 1 = 5$

   Departments S.S. can be partitioned into 2 components

   (a) practice S.S.

   $$= \frac{\Sigma t^2}{n_t} - C = \frac{604^2 + 541^2}{9} - C = 73\,055 \cdot 22 - 72\,834 \cdot 72$$

   $$= 220 \cdot 5$$

   Practice $df = k - 1 = 2 - 1 = 1$

   Practice var. est. $=$ S.S./$df = 220 \cdot 5/1 = 220 \cdot 5$

140

(b) Departments within practice groups S.S.

$$= \text{Departments S.S.} - \text{Practice S.S.}$$

$$= 651 \cdot 61 - 220 \cdot 50 = 431 \cdot 11$$

within sessions $df = $ Department $df - $ Practice df

$$= \qquad 5 \qquad - \qquad 1 \qquad = 4$$

within session var. est. $= 431 \cdot 11/4 = 107 \cdot 78$.

Note. The within practice groups var. est. is the proper error term for the practice groups var. est.

4. Reasoning ability (I.Q.) S.S.

$$= \frac{\Sigma t^2}{n_t} - C = \frac{470^2 + 359^2 + 316^2}{6} - C = 74\,939 \cdot 50 - 72\,834 \cdot 72$$

$$= 2104 \cdot 78$$

I.Q. $df = k - 1 = 3 - 1 = 2$

I.Q. var. est. $= 2104 \cdot 78/2 = 1052 \cdot 39$

5. Reasoning ability (I.Q.) × Practice interaction

Totals		High	Medium	Low	
	With	247	175	182	
	Without	223	184	134	$n_t = 3$

Interaction S.S. $= \dfrac{\Sigma t^2}{n_t} - \text{I.Q.S.S.} - \text{Practice S.S.} - C$

$$= \frac{247^2 + 175^2 + \ldots + 134^2}{3} - 2104 \cdot 78 - 220 \cdot 50 - 72\,834 \cdot 72$$

$$= 273 \cdot 00$$

Interaction $df = $ I.Q. $df \times $ Practice $df = 2 \times 1 = 2$

Interaction var. est. $= $ S.S./$df = 273 \cdot 00/2 = 136 \cdot 5$

6. Residual S.S. $= $ Total S.S. $- $ Department S.S. $- $ I.Q.S.S. $- $ I.Q. × Practice interaction

$$= 3368 \cdot 28 - 651 \cdot 61 - 2104 \cdot 78 - 273 \cdot 00 = 338 \cdot 89$$

Summary table

Source	S.S.	df	var. est.	F	df_1, df_2	Significance
(Departments	651·61	5)				
Practice groups	220·50	1	220·50	2·05	1, 4	NS
Departments within practice groups (error a)	431·11	4	107·78			
I.Q.	2104·78	2	1052·39	24·84	2, 8	<0·1%
I.Q. × Practice	273·00	2	136·50	3·22	2, 8	NS
Residual (error b)	338·89	8	42·36			
Total	3368·28	17				

Conclusion. The difference between students who do and do not have supplementary practice sessions is not statistically significant. It should be noted that this test is not very sensitive because of the small number of degrees of freedom (4) associated with the error term used to test the practice effect.

7.5   Trends and comparisons testing

Trend tests and comparisons (both planned and unplanned) between groups use the same set of techniques which are briefly summarized here. Both use coefficients (sometimes called lambda coefficients) to weight group totals:

$$
\begin{array}{llllll}
\text{Totals} & t_1 & t_2 & \cdots & t_k & \\
\text{Coefficients} & c_1 & c_2 & \cdots & c_k & : \quad \Sigma c = 0 \\
\text{Cross product} & c_1 t_1 & c_2 t_2 & \cdots & c_k t_k &
\end{array}
$$

$$\text{Sum of squares} = \frac{(\Sigma ct)^2}{n_t \Sigma c^2} \tag{7.5.1}$$

where n_t is the number of scores which have contributed to each total. This formula assumes that n_t is the same for each total.

The following points should be noted:

Degrees of freedom are always 1 for each sum of squares.

A set of coefficients must always *sum to zero*.

For planned comparisons, many of the coefficients are zero.

When more than one trend or *planned* comparison is used the coefficients are restricted by the requirement that all sets of coefficients used must be *mutually orthogonal*. Two sets of coefficients are said to be mutually orthogonal when their cross product sum is zero. For example:

Set 1	-2	-1	0	1	2	: Sum $= 0$
Set 2	1	-4	6	-4	1	: Sum $= 0$
Cross product	-2	$+4$	0	-4	$+2$	: Sum $= 0$

These two sets are mutually orthogonal. Unplanned comparisons are not subject to this restriction. Orthogonal coefficients suitable for use in trend analysis are given in table 7.5.

Table 7.5   Orthogonal coefficients for analysis of variance trend tests.

Number of groups	Trend	Coefficients						
3	Linear	-1	0	1				
	Quadratic	1	-2	1				
4	Linear	-3	-1	1	3			
	Quadratic	1	-1	-1	1			
	Cubic	-1	3	-3	1			
5	Linear	-2	-1	0	1	2		
	Quadratic	-2	1	2	1	-2		
	Cubic	-1	2	0	-2	1		
	Quartic	1	-4	6	-4	1		
6	Linear	-5	-3	-1	1	3	5	
	Quadratic	5	-1	-4	-4	-1	5	
	Cubic	-5	7	4	-4	-7	5	
	Quartic	1	-3	2	2	-3	1	
7	Linear	-3	-2	-1	0	1	2	3
	Quadratic	5	0	-3	-4	-3	0	5
	Cubic	-1	1	1	0	-1	-1	1
	Quartic	3	-7	1	6	1	-7	3

Unplanned comparisons differ in that the obtained F value for the comparison must be divided by $(k-1)$ *before the tables are consulted*:

$$F' = F/(k-1) \tag{7.5.2}$$

where

F' has $(k-1)$ and $N-k$ degrees of freedom
k is the number of available group totals
N is the total number of scores.

7.5.1 Trend test example

A group of 20 students is subdivided into 4 groups of differing reasoning ability. These range from 1 (low ability) to 4 (high ability). Their examination scores are given below along with a test for three kinds of trend. Only the linear trend is of interest but the others are given for the purposes of illustration.

Reasoning ability

1	2	3	4	
37	36	57	51	
43	46	59	69	$n_t = 4$
47	53	63	71	$N = 16$
51	61	72	73	
178	196	251	264	$T = 889$

$$\Sigma X^2 = 51\,585$$

1. $C = T^2/N = 889^2/16 = 49\,395$

2. Total S.S. $= \Sigma X^2 - C = 51\,585 - 49\,395 = 2190$

   Total $df = N - 1 = 16 - 1 = 15$

3. Between groups S.S. $= \dfrac{\Sigma t^2}{n_t} - C = \dfrac{178^2 + 196^2 + 251^2 + 264^2}{4} - 49\,395 = 1304 \cdot 2$

   Between groups $df = k - 1 = 4 - 1 = 3$

3a. Linear S.S.

t	178	196	251	264	
c	-3	-1	1	3	
ct	-534	-196	$+251$	$+792$	$\Sigma ct = 313$

$$\Sigma c^2 = (-3)^2 + (-1)^2 + (1)^2 + (3)^2 = 20$$

Linear S.S. $= \dfrac{(\Sigma ct)^2}{n_t \Sigma c^2} = \dfrac{313^2}{4 \times 20} = 1224 \cdot 6$

3b. Quadratic S.S.

t	178	196	251	264	
c	1	-1	-1	1	
ct	178	-196	-251	264	$\Sigma ct = -5$

$$\Sigma c^2 = 1^2 + (-1)^2 + (-1)^2 + 1^2 = 4$$

Quadratic S.S. $= \dfrac{(\Sigma ct)^2}{n_t \Sigma c^2} = \dfrac{-5^2}{4 \times 4} = 1 \cdot 6$

3c. Cubic S.S.

t	178	196	251	264	
c	-1	3	-3	1	
ct	-178	588	-753	264	$\Sigma ct = -79$

$$\Sigma c^2 = (-1)^2 + 3^2 + (-3)^2 + 1^2 = 20$$

Cubic S.S. $= \dfrac{(\Sigma ct)^2}{n_t \Sigma c^2} = \dfrac{(-79)^2}{4 \times 20} = 78 \cdot 0$

4. Within groups S.S.

$$\text{W.G. S.S.} = \text{total S.S.} - \text{B.G.S.S.} = 2190 - 1304 \cdot 18$$

$$= 885 \cdot 82$$

$$\text{W.G. } df = \text{total } df - \text{B.G. } df = 15 - 3 = 12$$

Summary table

Source	S.S.	df	var. est.	F	df_1, df_2	Significance
Between groups	1304·2	3	434·7	5·89	3, 12	<2·5%
(a) linear	1224·6	1	1224·6	16·59	1, 12	<1%
(b) quadratic	1·6	1	1·6	<1	1, 12	NS
(c) cubic	78·0	1	78·0	1·1	1, 12	NS
Within groups	885·8	12	73·8			
Total	2190	15				

Conclusion.   There is a significant linear trend relating ability with examination performance.

7.5.2. Planned comparison example

Students from five departments were asked to estimate the number of hours spent in private study of statistics per week. The answers (in hours) are given below. The following comparisons were planned; (i) Sociology vs. Economics; (ii) (Botany, Zoology and Psychology) vs. (Sociology and Economics); (iii) Psychology vs. (Botany and Zoology).

Botany	Zoology	Psychology	Sociology	Economics	
2·4	0·7	2·4	0·3	0·5	$n_t = 4$
2·7	1·6	3·1	0·3	0·9	$N = 20$
3·1	1·7	5·4	2·4	1·4	$k = 5$
3·1	1·8	6·1	2·7	2·0	$T = 44·6$
11·3	5·8	17·0	5·7	4·8	

1. $C = T^2/N = 44·6^2/20 = 99·46$

2. $\text{T.S.S.} = \Sigma X^2 - C = 143·44 - 99·46 = 43·98$

   Total $df = N - 1 = 20 - 1 = 19$

3. Between groups

$$\text{B.G. S.S.} = \frac{\Sigma t^2}{n_t} - C = \frac{505·9}{4} - 99·46 = 27·0$$

$$\text{B.G. } df = k - 1 = 5 - 1 = 4$$

Comparison 1 Sociology vs. Economics

t	11·3	5·8	17·0	5·7	4·8	
c	0	0	0	+1	−1	
ct	0	0	0	5·7	−4·8	$\Sigma ct = 0·9$

$$\Sigma c^2 = (+1)^2 + (-1)^2 = 2$$

$$\text{Comparison S.S.} = \frac{(\Sigma ct)^2}{n_t \Sigma c^2} = \frac{0·9^2}{4 \times 2} = 0·10$$

Comparison 2 (Botany, Zoology and Psychology) vs. (Sociology and Economics)

t	11·3	5·8	17·0	5·7	4·8	
c	+2	+2	+2	−3	−3	
ct	22·6	11·6	34·0	−17·1	−14·4	$\Sigma ct = 36·7$

$$\Sigma c^2 = 2^2 + 2^2 + 2^2 + (-3)^2 + (-3)^2 = 30$$

$$\text{Comparison S.S.} = \frac{(\Sigma ct)^2}{n_t \Sigma c^2} = \frac{36·7^2}{4 \times 30} = 11·22$$

Comparison 3 (Botany and Zoology) vs. Psychology

t	11·3	5·8	17·0	5·7	4·8	
c	+1	+1	−2	0	0	
ct	11·3	5·8	−34·0	0	0	$\Sigma ct = -16·9$

$$\Sigma c^2 = 1^2 + 1^2 + (-2)^2 = 6$$

$$\text{Comparison S.S.} = \frac{(\Sigma ct)^2}{n_t \Sigma c^2} = \frac{-16·9^2}{4 \times 6} = 11·9$$

4. Within groups

$$\text{W.G. S.S.} = \text{T.S.S.} - \text{B.G.S.S.} = 43\cdot98 - 27 = 16\cdot98$$

$$\text{W.G. } df = \text{T } df - \text{B.G. } df = 19 - 4 = 15$$

Summary table

Source	S.S.	df	var. est.	F	df_1, df_2	Significance
Between groups	27·0	4	6·75	6·0	4, 15	<1%
Comparison 1	0·11	1	0·11	0·1	1, 15	NS
2	11·22	1	11·22	9·9	1, 15	<1%
3	11·90	1	11·90	10·5	1, 15	<1%
Within groups	16·98	15	1·13			
Total	43·98	19				

Conclusion. The departments differ in the amount of time students spend in the study of statistics. In particular, the laboratory based sciences seem to spend more time than the Sociology and Economics students. Within the laboratory based sciences, Psychology students spend most time on the topic.

7.5.3 Unplanned comparison example

The difference between Psychology and Sociology can be used to illustrate the procedure for an unplanned comparison.

t	11·3	5·8	17·0	5·7	4·8	
c	0	0	+1	−1	0	
ct	0	0	17·0	−5·7	0	$\Sigma ct = 11\cdot3$

$$\Sigma c^2 = 1^2 + (-1)^2 = 2$$

$$\text{Comparison S.S.} = \frac{(\Sigma ct)^2}{n_t \Sigma c^2} = \frac{11\cdot3^2}{4 \times 2} = 15\cdot96$$

Comparison $df = 1$

Comparison var. est. = S.S./df = 15·97/1 = 14·97

$$F = \frac{\text{Comparison var. est.}}{\text{Within groups var. est.}} = \frac{15\cdot97}{1\cdot13} = 14\cdot13$$

For *unplanned* comparisons compute F^1

$$F^1 = F/(k-1) = 14\cdot13/(5-1) = 14\cdot13/4 = 3\cdot5$$

$$F^1 \text{ has } (k-1) \text{ and } (N-k) \text{ } df$$

$$\text{has} \quad 4 \quad \text{and} \quad 15 \quad df$$

This value is significant at the 5 per cent level.

8. Product moment correlation

8.1 Pearson's *r*

We use Pearson's r to describe the degree of linear correlation between two variables X and Y. Usually, we have a sample of N pairs (X, Y) of scores and we wish to decide whether *in general* we can make useful predictions of the value of Y when we know the corresponding X score.

We compute r thus

$$r = \frac{N\Sigma XY - \Sigma X\,\Sigma Y}{\sqrt{[(N\Sigma X^2 - (\Sigma X)^2)(N\Sigma Y^2 - (\Sigma Y)^2)]}} \tag{8.1}$$

where r is a measure of correlation whose value lies between -1 and $+1$. Values of r near to $0\cdot0$ indicate a low degree of correlation. Positive values of r indicate that high values of X are associated with high values of Y and *vice versa*. Negative values of r indicate that low values of X are associated with high values of Y. See 8.2 for worked example.

8.1.1 Testing the significance of *r*

The statistical significance of r can be assessed directly by consulting table 8.1. Since r may have either a positive or negative sign, a 1-TAIL test is appropriate if a correct and properly reasoned prediction of its sign was made before the results were available.

Table 8.1 describes the situation where the following assumptions are met:

  (i) X and Y are not correlated variables.
 (ii) The samples were drawn at random.
(iii) The samples were drawn from normally distributed populations.

When these assumptions are met we expect our sample value of r to be approximately zero. A significantly large value of r (large positive or large negative) may indicate that one of these assumptions is not met. Proper experimental procedure should guarantee assumption (ii) and the test should not be used if assumption (iii) is not valid. This leaves the possibility that the X and Y scores are correlated and that particular values of X are associated with particular values of Y. Moreover, we know that the value of r will be greatest when the relationship is linear.

The significance of r may also be investigated by computing t:

$$t_{N-2} = \frac{r}{\sqrt{[(1 - r^2)/(N - 2)]}} \tag{8.2}$$

where t has $N - 2$ degrees of freedom and may be assessed by reference to table 2.7.5. t takes the sign of r and the above considerations of 1- and 2-TAIL tests still apply.

Table 8.1   Critical values of *r*, Pearson's correlation coefficient.

2-TAIL	10%	5%	2%	0·2%
1-TAIL	5%	2·5%	1%	0·1%
$N = 4$	$r \geq 0.90$	0·95	0·98	0·995
5	0·81	0·88	0·93	0·97
6	0·73	0·81	0·88	0·94
7	0·67	0·75	0·83	0·90
8	0·62	0·71	0·79	0·87
9	0·58	0·67	0·75	0·84
10	0·55	0·63	0·72	0·81
11	0·52	0·60	0·69	0·78
12	0·50	0·58	0·66	0·76
13	0·48	0·55	0·63	0·74
14	0·46	0·53	0·61	0·72
15	0·44	0·51	0·59	0·70
16	0·43	0·50	0·57	0·68
17	0·41	0·48	0·56	0·66
18	0·40	0·47	0·54	0·65
19	0·39	0·46	0·53	0·64
20	0·38	0·44	0·52	0·62
21	0·37	0·43	0·50	0·61
22	0·36	0·42	0·49	0·60
23	0·35	0·41	0·48	0·59
24	0·34	0·40	0·47	0·57
25	0·34	0·40	0·46	0·56
30	0·31	0·36	0·42	0·52
40	0·26	0·30	0·36	0·45
50	0·24	0·28	0·34	0·42
60	0·22	0·26	0·31	0·38
70	0·20	0·24	0·28	0·35
80	0·18	0·22	0·26	0·33
90	0·17	0·21	0·24	0·32
100	0·16	0·20	0·23	0·30

8.2 Regression

We can describe the association of X and Y in terms of the best fit straight line graph:

$$Y' = bX + C \qquad (8.2.1)$$

where:

X is an observed score
Y' is a predicted value of Y
b and C are constants.

This line is chosen so that the errors of prediction $(Y - Y')$ are as small as possible. To be specific:

$$\text{Error}: \Sigma(Y - Y')^2 = \text{Minimum}$$

With this criterion, we can specify formulae for calculating b and C;

$$b = \frac{N\Sigma XY - \Sigma X\Sigma Y}{N\Sigma X^2 - (\Sigma X)^2} \qquad (8.2.2)$$

$$C = \frac{\Sigma Y - b\Sigma X}{N} \qquad (8.2.3)$$

A best fit line is shown in Fig. 8.2. The term b in formula (8.2.2) refers to the *slope* of the best fit line. In fact:

$$b = \tan \alpha°$$

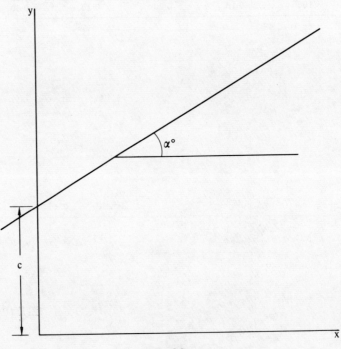

Fig. 8.2.

The term C in formula (8.2.3) refers to the height above the X axis where the best fit line cuts the Y axis. Regression equations are usually only meaningful if the correlation (r) between the variables can be shown to be significant.

▶ Example. Twenty students about to begin an introductory course in statistics completed an abstract reasoning test. Their scores were given on a scale from 1–100. These and their examination results (percentage scale) at the end of the course are given below. Are the examination results correlated with the abstract reasoning test performance as predicted?

SCORES: ALL INDEPENDENT
VARIABLES: (a) REASONING TEST, POSSIBLY NORMALLY DISTRIBUTED
                   (b) EXAMINATION RESULTS, POSSIBLY NORMALLY DISTRIBUTED
DATA SUMMARY TABLE:

(X) Reasoning test	27 37 43 46 50 51 51 51 53 55 58 61 61 67 68 70 73 76 76 81
(Y) Examination result	43 27 18 23 47 37 63 47 64 55 74 43 64 54 54 80 67 69 57 96

1. $\Sigma X = 1155$; $\Sigma X^2 = 70\,461$

2. $\Sigma Y = 1082$; $\Sigma Y^2 = 65\,676$

3. $\Sigma XY = 66\,208$; $N = 20$

4.
$$r = \frac{N\Sigma XY - \Sigma X \Sigma Y}{\sqrt{(N\Sigma X^2 - (\Sigma X)^2)(N\Sigma Y^2 - (\Sigma Y)^2)}} = \frac{74\,450}{\sqrt{75\,195 \times 142\,796}} = 0.72$$

5. Consulting table 8.1 we find that our value of r is significant at better than the 0·1 per cent level (1-TAIL test)

6. The regression values are

$$b = \frac{N\Sigma XY - \Sigma X \Sigma Y}{N\Sigma X^2 - (\Sigma X)^2} = \frac{74\,450}{75\,195} = 0.99$$

$$C = \frac{\Sigma Y - b\Sigma X}{N} = -3\cdot1$$

$$Y' = 0.99X - 3\cdot1$$

Conclusion. The significant value of r (0·72) indicates a linear correlation between reasoning test results and examination results.

8.3 Difference between two sample values of Pearson's r

When we have two values of r obtained from two samples, we often wish to know whether they are significantly different.

Procedure

1. Calculate r_1 and r_2 from the data. The samples contain n_1 and n_2 scores, respectively.
2. Convert r_1 and r_2 to T_{r_1} and T_{r_2} (Fisher's transform) using table 8.3.
3. Calculate Z

$$Z = \frac{T_{r_1} - T_{r_2}}{\sqrt{\left(\dfrac{1}{n_1 - 3} + \dfrac{1}{n_2 - 3}\right)}} \tag{8.3}$$

4. The significance of Z may be assessed by referring to table 2.7.2. Some critical values of Z are reproduced below for convenience. A 1-TAIL test is appropriate if a correct and properly reasoned prediction was made, as to which value of r would be larger, before the results were available.

2-TAIL	10%	5.0%	2.0%	0.2%
1-TAIL	5%	2.5%	1.0%	0.1%
$Z \geq$	1.64	1.96	2.33	2.98

Inferences. This test is based on the following assumptions:

(i) The samples were drawn from two populations which show the same correlation between the two variables in question.
(ii) The two samples were drawn independently of each other.
(iii) The two samples were drawn at random from normally distributed populations.

A significantly large value of Z may indicate that one of these assumptions is not true. The test should only be used when assumptions (ii) and (iii) are true. This leaves the possibility that the two samples were *not* drawn from populations where the correlation between our two variables is the same.

▶ Example. On two successive years we calculate a sample correlation between abstract reasoning ability and examination result. In the first year, we obtain a correlation of 0.63 with a sample of 64 students. In the second year, we obtain a correlation of 0.41 with a sample of 63 students. Can we say that reasoning ability was not so important a determinant of examination score for students on the course for the second year?

1. $r_1 = 0.63 \qquad r_2 = 0.41 \qquad n_1 = 64 \qquad n_2 = 63$

2. $T_{r_1} = 0.74 \qquad T_{r_2} = 0.44$

3. Using formula (8.3)

$$Z = \frac{0.74 - 0.44}{\sqrt{(1/61 + 1/60)}} = \frac{0.3}{\sqrt{(0.0164 + 0.0167)}}$$

$$= \frac{3}{0.18} = 16.5$$

4. Our value of Z is significant at better than the 0.2 per cent significance level on a 2-TAIL test (no prediction made).

Conclusion. The correlation between abstract reasoning ability and examination result was significantly lower on the latter of the two years. Reasoning ability was a less important determinant of success in the second year.

Table 8.2 Fisher's transform of Pearson's product moment correlation coefficient.

r	T_r	r	T_r
0·01	0·0100	0·51	0·5627
0·02	0·0200	0·52	0·5763
0·03	0·0300	0·53	0·5901
0·04	0·0400	0·54	0·6042
0·05	0·0500	0·55	0·6184
0·06	0·0601	0·56	0·6328
0·07	0·0701	0·57	0·6475
0·08	0·0802	0·58	0·6625
0·09	0·0902	0·59	0·6777
0·10	0·1003	0·60	0·6931
0·11	0·1104	0·61	0·7089
0·12	0·1206	0·62	0·7250
0·13	0·1307	0·63	0·7414
0·14	0·1409	0·64	0·7482
0·15	0·1511	0·65	0·7753
0·16	0·1614	0·66	0·7928
0·17	0·1717	0·67	0·8107
0·18	0·1820	0·68	0·8291
0·19	0·1923	0·69	0·8480
0·20	0·2027	0·70	0·8673
0·21	0·2132	0·71	0·8872
0·22	0·2237	0·72	0·9076
0·23	0·2342	0·73	0·9287
0·24	0·2448	0·74	0·9505
0·25	0·2554	0·75	0·9730
0·26	0·2661	0·76	0·9962
0·27	0·2769	0·77	1·0203
0·28	0·2877	0·78	1·0454
0·29	0·2986	0·79	1·0714
0·30	0·3095	0·80	1·0986
0·31	0·3205	0·81	1·1270
0·32	0·3316	0·82	1·1568
0·33	0·3428	0·83	1·1881
0·34	0·3541	0·84	1·2212
0·35	0·3654	0·85	1·2562
0·36	0·3769	0·86	1·2933
0·37	0·3884	0·87	1·3331
0·38	0·4001	0·88	1·3758
0·39	0·4118	0·89	1·4219
0·40	0·4236	0·90	1·4722
0·41	0·4356	0·91	1·5275
0·42	0·4477	0·92	1·5890
0·43	0·4599	0·93	1·6584
0·44	0·4722	0·94	1·7380
0·45	0·4847	0·95	1·8318
0·46	0·4973	0·96	1·9459
0·47	0·5101	0·97	2·0923
0·48	0·5230	0·98	2·2976
0·49	0·5361	0·99	2·6467

Appendix A. Random numbers

Table A1 contains 2500 randomly selected digits. This table was constructed using a computerized random number generator. You can use these tables to help you carry out a wide range of tasks which require a random element. These often involve a certain amount of ingenuity on your part. An example given on the following page shows how the tables can be used.

It is very important that you should not use the same sequence of random numbers each time, otherwise they will cease to be random. For this reason, the table has been subdivided into 100 blocks with 10 blocks per row. To choose a block at random close your eyes and touch the page with a pencil. Let the digit indicated by the pencil point be the number of the block in the row. Repeat the exercise and let your new digit indicate the row number. For example, if my pencil points to the digits 6 and 4, then I shall use the block:

$$
\begin{array}{ccccc}
4 & 2 & 4 & 7 & 9 \\
4 & 5 & 1 & 3 & 3 \\
5 & 7 & 7 & 9 & 4 \\
0 & 7 & 5 & 2 & 7 \\
5 & 0 & 5 & 2 & 0 \\
\end{array}
$$

for my purposes. You may now use the block from the top left hand corner or you may use additional techniques to decide exactly where in the block you will begin. For example, you could repeat the above exercise but use only digits 1–5 since there are only 5 rows and 5 columns per block.

▶ Example. Assign nine objects (A, B, C, D, E, F, G, H, I) at random to two groups so that both groups contain at least four objects.

I select my block of random numbers using the procedure suggested above. My pencil point chooses the digits 7 and 8 and I shall therefore use the block in column 7 and row 8:

$$
\begin{array}{ccccc}
5 & 3 & 4 & 8 & 0 \\
7 & 2 & 2 & 0 & 9 \\
8 & 7 & 9 & 4 & 8 \\
8 & 5 & 6 & 3 & 4 \\
5 & 5 & 2 & 5 & 7 \\
\end{array}
$$

I start from the top left hand corner, proceed from left to right and obtain the following random sequence of nine numbers:

$$5 \quad 3 \quad 4 \quad 8 \quad 0 \quad 7 \quad 2 \quad 2 \quad 0$$

I assign the objects A–I in order using the following rule. If the corresponding number is odd, I assign the object to group (i). If it is even (zero is even), I assign it to group (ii). If one group comes to contain five objects, then all the remaining objects are assigned to the other group irrespective of the value of the random number.

Object	Random value	Group
A	5	(i)
B	3	(i)
C	4	(ii)
D	8	(ii)
E	0	(ii)
F	7	(i)
G	2	(ii)
H	2	(ii)
I	(0)	(i)

Note: Object I is assigned to group (i) even though its random value (0) is even. This guarantees that group (i) contains at least four objects.

The final arrangement is as follow:

Group (i)   A, B, F, I
Group (ii)   C, D, E, G, H

Table A1   Random numbers.

	0	1	2	3	4	5	6	7	8	9
	03918	86495	47372	21870	28522	99445	38783	83307	10041	35095
	66357	64569	08993	20429	28569	63809	43537	58268	80237	17407
0	89680	04655	24678	61932	64301	47201	31905	60410	80101	33382
	95255	10353	43857	42186	77011	93839	28380	49296	63311	49713
	91823	39794	47046	78563	89328	39478	04123	19287	34017	87878
	35674	39212	98246	29735	09924	27893	49105	00755	39242	50472
	39581	44036	54518	46865	72479	02741	75732	99808	02382	77201
1	44932	88978	84281	45650	28016	77753	39495	41847	19634	82681
	61589	35486	59500	20050	89769	54870	75586	07853	25318	01995
	87789	41212	74907	90734	31946	24921	40113	37395	51406	98099
	43023	70195	07013	72306	58420	43526	15539	24845	15582	16780
	95286	69021	18075	45894	09875	42869	20618	07699	80671	54287
2	52754	73124	93276	71521	59618	44966	37502	15570	05255	53579
	08239	99174	75548	95776	42314	13093	76032	35569	28738	38092
	74669	00749	17832	64855	97050	31553	32350	51491	53659	89336
	36912	05292	29030	43074	84602	95131	22769	44680	68492	33987
	28124	29686	63745	12313	15745	11570	20953	17149	97469	41277
3	90524	36459	22178	63785	20466	67130	91754	40784	38916	12949
	76104	20556	34001	59133	84599	29798	57707	57392	91757	76994
	43826	69089	06490	42228	94940	10668	62072	58983	10263	08832
	30666	02218	89355	76117	75167	69005	42479	79865	87228	15736
	08506	29759	74257	85594	75154	48664	45133	49229	32502	99698
4	68202	44704	39191	73740	55713	98670	57794	64795	27102	83420
	26630	95009	20390	38266	30138	61250	07527	02014	43972	49370
	13400	68249	32459	41627	56194	93075	50520	96784	08900	87788
	73717	19287	69954	45917	80026	55598	86757	47905	16890	99047
	78249	93739	97076	00525	19862	54700	18777	22218	25414	13151
5	54954	80615	96282	11576	59837	27429	60015	40338	39435	94021
	17463	26715	71680	04853	55725	87792	99907	67156	44880	55285
	95472	57551	24602	98311	63293	58110	61911	78152	96341	31473
	58398	61602	38143	93833	07769	22819	58373	88466	71341	32772
	93643	92855	73063	63623	29388	89507	78553	62792	89343	27401
6	24187	60720	74055	36902	22047	09091	79368	35408	06875	53335
	91274	87824	04137	77579	54266	38762	23393	37710	46457	03553
	58275	11138	18521	59667	00980	73632	88008	10060	48563	31874
	90785	78923	46611	39359	98036	25351	88031	62020	13837	03121
	56644	79453	49072	30594	73185	81691	29225	70495	98350	36891
7	04873	71321	29929	37145	95906	41005	17444	61728	86112	76261
	92519	61569	65672	95772	45785	21301	89563	23018	60423	50801
	70564	45398	54369	08513	36838	19805	67827	74938	66946	01206
	01698	72899	42819	76248	01666	20536	33090	53480	11035	80190
	37894	91343	40681	03155	63778	03384	64869	72209	55902	49417
8	16170	66725	12066	03367	21859	16726	38599	87948	61585	01927
	65438	93196	53717	69188	23299	05798	87281	85634	80075	45250
	85814	33345	91245	29321	97745	99216	19271	55257	98238	77331
	30234	64553	17232	19953	73033	63001	08226	70272	63125	05544
	62864	43257	87727	73924	50171	26664	84888	71019	83994	58793
9	11005	22169	26611	88784	62205	94075	30740	52637	71213	86084
	14862	15592	81328	36747	89773	38872	81329	25227	96907	47854
	92885	18368	93102	21342	43283	03982	97421	09952	77272	62492

Appendix B. Bibliographical notes

Further reading

In a book of this kind, there is probably little to be gained by including a comprehensive list of original research papers which contain the material on which the tests are based. The style of this text was determined by the fact that most non-statisticians are ill equipped to tackle such papers. Accordingly, I shall restrict the reference list as far as possible to books and papers whose role is principally that of interpreter, of intermediary, between the statistical theorists and the student. These texts contain the required detailed references to the original work.

I have restricted the number of major sources to three—Edwards (1962), McNemar (1962), and Siegel (1956)—in an attempt to make the business of 'looking up' as stress-free as possible. Between them they cover almost the whole of the ground dealt with in this book. All three were written with psychologists in mind and the authors have, of necessity, gone to considerable trouble to make their exposition clear.

Failure to mention many worthy textbooks is not intended as any reflection upon them. The aim of this Bibliography is to present *at least one* readable reference.

Section 2.1–2.6 *Basic statistical procedures*

The material summarized here is covered in almost all introductory texts; McNemar (1962, pp. 1–37) is particularly clear.

2.7 *Standard frequency distributions*

More comprehensive tables than those given here can be found in *Biometrika Tables for Statisticians* (Pearson *et al.* 1958). McNemar (1962) discusses all of these distributions except Poisson.

3. *Choosing a test*

Siegel (1956, pp. 1–44) provides an excellent introduction to this area which leaves little to be desired. Winer (1971, pp. 10–14) deals concisely with many of the important terms.

4. *Tests for one variable*

Binomial, Chi square and K–S (ranked categories) tests are discussed in Siegel (1956). McNemar (1962), too, deals with chi-square tests. I am unaware of any simple description of the K–S test for rankable scores, but the reader may care to try Goodman (1954).

5. *Tests for two variables* (unrelated scores)

Siegel (1956) covers all of the tests in this section except for the unrelated t test and Kendall's τ for ranked categories. The t test is covered very well by Edwards (1962) and Winer (1971). Kendall (1948) is the best source for the ranked categories version of τ.

6. *Tests for two variables* (related scores)

The related t test is discussed by Edwards (1962) and Winer (1971). The McNemar test for changes is, of course, dealt with by McNemar (1962) but somewhat more clearly by Siegel (1956) who also presents Cochran's test, the sign test, Wilcoxon's test and Friedman's test. The two groups repeated measures test is given by Meddis (1975) and Page's L test by Page (1963).

7. *Analysis of variance*

By far the most useful book, here, from a practitioner's point of view is Edwards (1963), although Winer (1971) gives a more rigorous account of the theoretical issues. For beginners, Meddis (1973) is a self-teaching workbook.

8. *Product moment correlation*

McNemar (1962) covers this topic very adequately.

References

Edwards, A. L. 1962. *Experimental Design in Psychological Research.* New York: Holt, Rinehart and Winston.

Goodman, L. A. 1954. 'Kolmogorov–Smirnov tests for psychological research.' *Psychol. Bull.*, 51, 160–168.

Kendall, M. G. 1948. *Rank Correlation Methods.* London: Griffin.

McNemar, Q. 1962. *Psychological Statistics.* New York: Wiley.

Meddis, R. 1973. *Elementary Analysis of Variance for the Behavioural Sciences.* London: McGraw-Hill.

Meddis, R. 1975. 'A simple two group test for matched scores with unequal cell frequencies'. *British Journal of Psychology.*

Page, E. B. 1963. 'Ordered hypotheses for multiple treatments: a significance test for linear ranks.' *Amer. Stat. Ass. J.*, 216–230.

Pearson, E. S. and Hartley, H. O. 1958. *Biometrika Tables for Statisticians.* Cambridge University Press.

Siegel, S. 1956. *Nonparametric Statistics for the Behavioural Sciences.* New York: McGraw-Hill.

Winer, B. J. 1971. *Statistical Principles in Experimental Design.* New York: McGraw-Hill.

Index

161

Printed in Great Britain by J. W. Arrowsmith Ltd, Bristol, BS3 2NT.